Computational Bioengineering

Current Trends and Applications

Computational Bioengineering

Current Trends and Applications

editors

M. Cerrolaza
Universidad Central de Venezuela, Venezuela

M. Doblaré
Instituto de Investigación en Ingeniería de Aragón (I3A), España

G. Martínez
Universidad Central de Venezuela, Venezuela

B. Calvo
Instituto de Investigación en Ingeniería de Aragón (I3A), España

Imperial College Press

Published by

Imperial College Press
57 Shelton Street
Covent Garden
London WC2H 9HE

Distributed by

World Scientific Publishing Co. Pte. Ltd.
5 Toh Tuck Link, Singapore 596224
USA office: Suite 202, 1060 Main Street, River Edge, NJ 07661
UK office: 57 Shelton Street, Covent Garden, London WC2H 9HE

British Library Cataloguing-in-Publication Data
A catalogue record for this book is available from the British Library.

COMPUTATIONAL BIOENGINEERING
Current Trends and Applications

ISBN 1-86094-465-5

Printed in Singapore by World Scientific Printers (S) Pte Ltd

PREFACE

Computational Bioengineering is a relatively recent and emergent discipline. Although the Finite element Method appeared more than four decades ago, other numerical methods such as the Boundary Elements and Meshless methods have been successfully used to solve problems in engineering and applied sciences.

Thus, numerical modeling and computer simulation, ranging from volume modeling to Immersive Virtual Reality (IVR), have been also widely used to create "virtual worlds", allowing the designers to interact with them and to modify them as well. This is an essential feature when dealing with bioengineering design, because bioengineers face permanently "time-changing" and living models. As an example of the enormous complexity involved, it can be said that artificial prosthesis replacement strongly depends on a rational and accurate evaluation of stress concentrations arising in the couple bone prosthesis and that bone behavior is greatly affected by the biological loads from the living organism activities.

More recently, researchers have realized the need to go further into the human tissue behavior. Thus, mechanobiology has emerged with a large potential for the simulation of biological actions. Therefore, scientists are now modeling the actions and their associated responses, ranging from the organ scale, passing through biological tissue, down to individual cells and molecules. This generates a big challenge and expectation, specially when remembering the well-known axiomatic statement: *the closer you look at the real world, the fuzzier it becomes.*

This book is a significant contribution to the state-of-the-art in the field of Computational Bioengineering, from the need of a Living Human Database to Meshless methods in biomechanics, from computational mechanobiology to the evaluation of stresses in hip prosthesis replacement, from Lattice-Boltzmann methods for analyzing blood flow to the analysis of fluid movement in long bones, among other interesting topics treated herein. Relevant and well known world-wide experts in bioengineering have contributed to this book, thus giving it a unique style and a "fronting-edge" material for academic researchers, postgraduate students and design bioengineers as well as all those interested in getting a better understanding in such complex and fascinating human and living processes.

The editors greatly appreciate the effort, dedication and work of the authors. At the same time, the editors accept no responsibility for the opinions and concepts emitted by the authors as well as for any consequence derived from the use, interpretation or practice based on the material contained in this book.

Golden, Colorado
The Editors, January 2004

This book has been co-edited by:

Instituto de Investigación en Ingeniería de Aragón (Universidad de Zaragoza, España)
Oficina de Planificación del Sector Universitario (Venezuela)
Universidad Central de Venezuela (Venezuela)

CONTENTS

THE LIVING HUMAN PROJECT: THE GRAND CHALLENGE OF EUROPEAN BIOMECHANICS?

MARCO VICECONTI, FULVIA TADDEI, ALESSANDRO CHIARINI, ALBERTO LEARDINI AND DEBORA TESTI

Istituti Ortopedici Rizzoli, Bologna, Italy

The release, some years ago, of the Visible Human (VH) dataset made it possible, for the first time, to access anatomical information without compromises. This produced a significant momentum in many areas. However, after some time, it became clear that, while the dissection approach used in the VH project ensured extreme quality, it also lacked many aspects that other forms of data contain. These include in vivo data collection, multi-subject, gender, sex, and age variations, lack of connection with functional information, no pathology, etc. To put it simply - he is just one, and he is dead. We do not know how he breathed, walked, swallowed, digested, and how the anatomy ad the relative functional aspects vary from subject to subject. The VH data totally lacks multiplicity and functionality. Many research projects have been carried out in Europe over the last few years, some with the support of the European Commission, to try to circumvent some of these limitations. A basic aspect possessed by the VH project, and lacking in all these other projects, is completeness. The VH project relates ONLY to the normal anatomy of one human subject, and provides ALL the anatomical information for that subject. The other projects had wider objectives, but lacked completeness. Some focused only on pathological data, others only on the lower limb, others again only on the modelling of functional aspects. Because of the lack of the necessary critical mass, none has dared to search for completeness. The Living Human Project (LHP) intend to develop a worldwide, distributed repository of anatomo-functional data and of simulation algorithms, fully integrated into a seamless simulation environment and directly accessible by any researcher in the world. The objective is patient-specific bio-numerics (-mechanics, -electromagnetic, etc.) and image-processing (both for pre-processing & visualisation) for the complete human body, with integration of individual systems through hierarchical approaches at the algorithmic level and through middleware operating across distributed systems for Grid computing, using a semantic web to manage the information. The focus of the Grid approach is to provide services to medical or clinical users, removing any need for them to have to handle the details of the computing systems or simulation methods. The initiative is a very early stage. This paper provides some background and also some preliminary ideas on how to implement this 'Biomechanics Grand Challenge'. Particular attention is given to the description of the Multimod Application Framework, Open Source software for the visualisation and the processing of biomechanical data, and of the possible architecture of the Living Human Digital Library.

1   Introduction: The Story So Far

1.1   *Biomechanics in Europe and the BioNet project*

It is said that any truly new knowledge is created at the borders of established science fields. Biomechanics is a perfect example. We define biomechanics as the science that studies the action of forces on living bodies at any scale, from whole body down to cells.

Biomechanics is not a scientific discipline in a strict sense. Rather, it is a discipline made of other disciplines. It is defined at the border between mathematics, physics, chemistry and biology, but also between medicine and engineering. Within Biomechanics coexist basic research and applied research, empiricism and rigorous application of the scientific method. Biomechanics has found a disparate range of applications, such as clinical medicine, physiology, rehabilitation, sport performance and safety, ergonomic safety and comfort, forensic medicine, just to name but a few. It is distinguished by an intensive use of enormous collections of medical imaging data, computer models, experimental and clinical measurements, and large datasets generated by simulation codes. The time-dependent character of much of the data increases substantially the volume and complexity that has to be catered for. The Biomechanics community contains a wide spectrum of researchers, from the intense technophile to the severe technophobe. In addition, these researchers need to communicate the results of their studies to the most heterogeneous user community, made of clinicians, industries, and groups of citizens with special needs or interests.

In Europe the biomechanics research community is characterised, beside some notable exceptions, by a large number of small research units, strongly connected to their local context (i.e. the local hospital, or the local engineering department) but struggling to find their own space within the national research scenario. European Biomechanics still suffer form a severe fragmentation. There is a considerable separation between ergonomy biomechanics, solid mechanics biomechanics, and movement analysis biomechanics communities, just to name a few examples. Each of these communities has its own conferences, its own scientific societies, its own journals, etc. This severe fragmentation comes with a cost. Although biomechanics and biomedical engineering are among the disciplines that more closely fit into the general scopes pursued by the European Commission with the framework programs for research and technological development, he grant opportunities for biomechanics constantly decreased over the years. Also the societal visibility of the discipline Biomechanics is negligible. Most people do not know what biomechanics is or what it studies.

Starting from these observations a small consortium launched in 2001 the BioNet project, aimed to promote the networking and the collaboration inside European Biomechanics. Central to the project was the organisation of the BioNet event held in Brussels in 2002 and attended by more than 100 delegates from the most prestigious European research institutions. This event was prepared with a very innovative approach. Through the BIOMCH-L mailing list, more than 4000 researchers were invited to suggest controversial topics in biomechanics. This produced seven general topics, which were extensively discussed – again on BIOMCH-L – one by one, by the whole community. As the discussion focused, the organisers also invited colleagues known to have significant experience in the specific context, to provide contributions to the discussion and to help in structuring the resulting positions. The process continued until the end of the first day of the BioNet Event, during which many invited speakers contributed to the discussion with their presentations. On the second day of the Event, a different speaker briefly presented each statement. To make the discussion of the statements more effective, three workgroups were created. The smaller size of these groups was expected to foster active participation, especially by younger researchers, and to promote a more focused discussion. Workgroup A collected all statements relating to the inherent nature of biomechanical modelling. Workgroup B discussed all statements relating to the sharing of research results. Workgroup C dealt with the statements relating to specific research contexts, in particular ergonomics, safety and biomaterials-tissue engineering. For each workgroup, a rapporteur provided a brief summary of the consensual conclusions to the subsequent plenary session. The final addresses of John O'Connor (University of Oxford, UK) and Steven Stanhope (National Institute of Health, USA) provided, respectively, a European and an American perspective on the work done. After the event, for each main topic (modelling in the support of physical mobility, sharing research data, biomechanics in ergonomics and safety, biomechanics in biomaterials and tissue-cell engineering) the BioNet consortium created a web-based discussion forum. These forums were used to continue the discussion, and to collectively revise the final consensus documents of the event. Following long discussion, the BioNet_VRLab Workgroup found consensus over the following points:

- sharing research results is, in general, a valuable process that can make our research results more relevant for humanity
- in most cases, the sharing of research results is expected to provide a positive effect on biomedical research
- public research-funding agencies and charities should require that results produced with their money are made publicly available, unless specific reasons make it impossible or inappropriate, but only if they can provide funding for the sharing activities
- the approaches adopted so far to share biomechanics research results have not been effective in promoting a direct or indirect commercial exploitation of the research results, nor even for collecting sufficient revenues to maintain the sharing effort
- a more structured approach to sharing is necessary, in order to exploit its value to the maximum, as well as to promote the sharing culture inside our community
- the creation of a global community would make the sharing of biomechanics research results much more useful and effective
- sharing initiatives should be organised under the common umbrella of a virtual community, which would be responsible for setting basic rules and procedures, monitoring processes, and fostering the creation of useful products for industrial, clinical or societal users

- a global sharing community would provide a useful structure to co-ordinate research results in answer to clinical, industrial and societal needs
- the sharing of research results would become more effective if it could rely on a distributed infrastructure supporting micro-payments, automatic accounting and billing, and policies and values that responded dynamically to changes in circumstances
- our scientific community should find a way to monitor the appropriateness of sharing initiatives, achieving an appropriate balance between control and openness
- a virtual community would provide a suitable structure for managing collective knowledge and for producing consensus statements, especially if innovative technology is available to support this process.
- The BioNet_VRLab Workgroup identified various challenges in achieving the goals above:
  - establishing a body of rules for the management of the community itself
  - ensuring the long-term sustainability of the operation
  - creating of a distributed infrastructure able to support the community, its internal marketplace, its consensus processes, and its knowledge management needs
  - coping with the technical difficulties of storing and sharing complex and highly heterogeneous information
  - promoting a direct or indirect commercial exploitation of research results, or simply the collection of the revenues required to maintain the sharing effort

The BioNet_VRLab Workgroup agreed that the creation of an Internet-based Biomechanics Research Community would dramatically improve the level of collaboration between research groups, with a clear synergistic effect on how we use research funding. In order to pursue these challenges, this Biomechanics Research Community needs a state-of-the-art technological infrastructure to support the information management as well as the applicability and the accountability of highly heterogeneous sharing policies.

1.2   The Biomechanics European Laboratory

As an immediate follow-up of these recommendations the BioNet consortium started to promote the creation of a virtual laboratory called Biomechanics European Laboratory (BEL). On September 2002 it was launched the VRLAB web site, aimed to promote the creation of the Biomechanics European Laboratory, a virtual laboratory that relying on a Grid infrastructure would allow all Biomechanics researchers to work collaboratively in a seamless way.

http://www.tecno.ior.it/VRLAB/BEL_home_page.html

Since then nearly 200 researchers joined the BEL. Using an Internet-based poll, the community appointed three delegates early after its foundation.

Another idea germinated in that period: to start a collective research project that would exploit at the highest possible level the collaborative opportunities provided by the BEL community: the Living Human Project.

2   The Living Human Project

2.1   Description of the Project

The Living Human Project (LHP) aims to develop a worldwide, distributed repository of anatomo-functional data and of simulation algorithms, fully integrated into a seamless simulation environment and directly accessible by any researcher in the world. The objective is patient-specific bio-numerics (-mechanics, -electromagnetic, etc.) and image-processing (both for pre-processing & visualisation) for the complete human body, with integration of individual systems through hierarchical approaches at the algorithmic level and through middleware operating across distributed systems for Grid computing, using a semantic web to manage the information. The focus of the Grid approach is to provide services to medical or clinical users, removing any need for them to have to handle the details of the computing systems or simulation methods.

LHP aims to create a complex combination of large data collections, sophisticated user interfaces, state-of-the-art simulation software environments, and of GRID-based distributed computational and storage infrastructures. All these components are necessary to create a generalised digital model of the human body functional anatomy. The Living human model would let researchers to:

- Access the totality of anatomical information provided by the Visible Human datasets in a format that could be directly used for biomechanics research.
- Replace a portion (an organ, a bone, a limb) of the generic model with subject-specific data, with all the tools required to scale the rest of the generic anatomical information to the subject-specific data.
- Create collections of multiple instances of any portion of the generic model, with all tools required to use this multiplicity in simulation studies to account for inter-subject variability.
- Associate to anatomical data collections of measurements on tissues material properties, movement data, postural data, motor control measurements, etc.
- Use all these data in combination with all the services provided by the simulation environments to create functional simulations, which can be easily customised by using some subject-specific data, or multiple-subject data to run statistical analyses.

- Combine these functional simulations with effective user interfaces to build pre-packaged solutions for specific medical problems.

This very rudimental and somehow naive description of such grand challenge already allowed the identification of some major milestones. A first long list of milestones regards all the information technologies we need to develop and implement:

- Establish the IT infrastructure required to set-up the virtual Laboratory.
- Establish the IT infrastructure required to form, to manage, to make accessible, and continuously update large biomedical data collections.
- Develop a multi-data registration toolkit allowing the averaging of similar data, as well as the merging of subject-specific data with complementary generic or average data.
- Develop a distributed software library to deploy Internet-based multimodal interfaces, customisable toward the application context.
- Develop a scriptable integration middleware providing support for pre-processing the data in the collection (i.e. meshing), support automatic code-specific input formatting, transparent access to commercial and research simulation codes with support for an Application Service Provider model, and automatic code-specific results formatting.
- Creation of a Grid-based computational and storage distributed infrastructure supporting all previously listed services, able to serve efficiently all Europe with an extensible deployment model that makes easy to add new nodes as new computing facilities decide to join the initiative.

At the same time it is necessary that the researchers in biomechanics and related topics are able to achieve these milestones:

- Creating an Internet biomechanics community fully representative of the various branches of the discipline as well as of the various countries that form, or will form in a near future, Europe.
- Organising training and re-training events on biomechanical modelling, the Living Human Project and on the use of the associated technologies.
- Creating awareness on the initiative and establishing communication with the neighbouring disciplines (physiology, anatomy, ergonomy, etc.) and with the natural 'customers': the clinicians, the industries, and the citizens at large.
- Form a sufficiently large (numerous and disparate) data collection.
- Process the anatomical information provided by the Visible Human datasets so to make it directly usable for biomechanics research.
- Extend and complement the Visible Human data collection with functional, pathological and multi-subject data.
- Associate to anatomical data collection of measurements on tissues material properties, movement data, postural data, motor control measurements, etc.
- Develop a systematic organisation of all the possible data (biomechanics and functional anatomy ontology).

- Create functional simulations easily customisable with subject-specific data, or multiple-subject data to run statistical analyses. Combine these functional simulations with effective user interfaces to build pre-packaged solutions for specific medical problems.

2.2 Implementing the Living Human Project

As it is formulated, the LHP is a mere theoretical exercise. In order to translate this Grand Challenge into practical milestones it is necessary the active involvement of many researchers, with different backgrounds and different application targets. In the following sections we shall describe two initiatives that if fully realised would provide a good part of what is necessary to translate the visionary idea of the Living Human Project into reality. Other similar projects are probably running right now, and more are expected to start in the next few years. If we shall be able to co-ordinate all these initiatives in a synergistic way, the vision may come true.

The first initiative is the result of a running research project supported by the European Commission in the frame of the FP5 IST action, entitled "Multimod: Simulation of multiple medical-imaging modalities: a new paradigm for virtual representation of musculo-skeletal structures". The Multimod Consortium, formed by CINECA, Istituti Ortopedici Rizzoli, University of Luton, Université Libre de Bruxelles, ESI Group and Vicon Ltd., has recently decided to release in the Public Domain for no-profit activities the major part of the software developed during the project. The Consortium will launch in late 2003 an Open Source Program called **Multimod Application Framework** (MAF). The MAF is a complex programming framework that allows an expert programmer to create a vertical application in the areas of computer-aided medicine and of computational biomechanics. Together with the MAF, the Multimod Consortium will also distribute an application called Multimod Data Manager. This program, intended for research purposes, will make available to the user all functions of the MAF, including all the multimodal visualisation capabilities the framework supports. In addition, the Data Manager will become the application of choice to read and write the Multimod Storage Format, a generalised file format that supports any type of data used in biomechanics, from imaging datasets, to motion analysis data, to finite element models. With its support for import and export functions toward the most popular file formats, the Multimod Storage Format is candidate to become the de facto standard to store and exchange biomechanics data.

The second initiative is currently only defined at a conceptual level. The project for realising the **Living Human Digital Library** was submitted to the European Commission for evaluation under the first call of the Sixth Framework Program. The idea behind the project is to create a new type of digital library services that would at the same time support the formation of the virtual community of European biomechanics, to make easier for the members of such community to share data and programs. The proposal received the third best score in its group, but is currently in a reserve list because of insufficient budget. Thus, this infrastructure project is not going to start any time soon. Nevertheless, we believe it is important to describe it in its essential lines, in the hope that sooner or later we shall be able to collect the resources required to realise this important infrastructure.

3   The Multimod Application Framework

3.1   General Description of the Framework

The Multimod Application Framework (hereinafter MAF) is a very ambitious software development project. The final framework should provide three layered software libraries that different users will exploit to realise the most disparate software systems.

A first possible mode of use is that of a programmer who wants to use some of the algorithms and functions we shall develop during the Multimod project in other applications, or to extend the MAF with its own algorithms. This use suggests a software layer independent for the GUI library, and easy to call from high-level programming languages.

The second use is that of an Application Expert who wants to develop a specialised computer aided medicine application targeting a specific medical diagnostic, treatment or rehabilitation service. This user wants to keep focus on the application and do as little as possible in terms of implementation. Ideally there should be a toolbox from which the user can pick only the operations required, specialise such operations by fixing parameters or by combing operations in specialised macro, and add an application-specific user interface.

The third mode of use is that of the final user. Here it is necessary to separate between the medical technician, who usually prepares the data, and the medical professional, who want to focus on the specific medical service to be performed. The technician usually has some understanding of the technical issues involved with the elaboration of digital data, and wants a powerful and productive user interface for the process of preparing and organising the data. The medical professional instead requires a user interface as familiar as possible, so he or she can focus on the task rather than on the application.

From these definitions of the scope and of the uses of the MAF, it is possible to derive a four-layer framework (Fig. 1).

The Multimod Consortium tried to avoid reinventing the wheel whenever possible. To fulfil its ambitious objectives, the MAF had to provide very sophisticated visualisation and processing of large scientific datasets. This was realised by basing the MAF on some of the best Open Source software libraries currently available. All visualisation services are obtained from the *Visualisation Tool Kit* (VTK). This library is becoming a sort of standard *de facto* in the area of

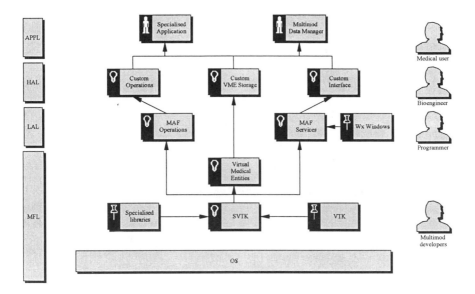

Figure 1. General Architecture of the Multimod Application Framework. For each architectural layer (left side) it is reported the type of user that is likely to exploit it (right side).

scientific visualisation, thanks to its clever architecture and to the powerful visualisation functions it provides [1]. VTK is the true foundation of the MAF; its architecture of classes has been adopted also for our *Multimod Foundation Layer* (MFL), the lowest of the three architectural layers forming the MAF. In this sense, the MFL may be seen as an extension of VTK. Because of this, all the classes we added to VTK to realise the MFL are grouped in a library called *Surgical-VTK* or SVTK for short. Thus, the two main elements forming the MAF foundation layer are VTK and SVTK.

To this core, we are adding specialised libraries as the need arise. The MAF already integrates the NLM Insight Tool-Kit (ITK) a new library developed by a prestigious consortium of six USA research institutions [2]. ITK implements the current state of the art for segmentation and registration algorithms of diagnostic datasets. Another specialised library that has been already integrated is V-Collide, a collision-detection package based on the OBB method [3].

In general these additional libraries are not fully integrated, but rather wrapped into a SVTK class that exposes the sub-set of methods useful to the MAF. Using these foundation libraries we developed a very powerful representation of the biomedical data, called *Virtual Medical Entity* (VME). Thanks to this sophisticated data structure, the MAF can integrate into a single program images, volumes, surfaces, motion data, movies, finite element meshes, etc.; MAF programs can cope with multidimensional time-varying datasets without any particular programming overhead.

The second layer of the architecture is called *Low Abstraction Layer*. It implements all *Virtual Medical Services*: the framework services such as undo or copy and paste, the *Operations*, high-level procedures that modify an existing VME, and the *Views*, which provide an interactive visualisation of the VMEs. The user interface is realised by integrating the MAF a portable GUI library, called WxWindows [4]. Thanks to this solution any application based on the MAF can be compiled on a variety of platforms. We are currently maintaining Windows and Linux distributions, but for other projects we have also compiled MAF programs under SGI Irix and other Unix operative systems. Although we never tried, in principle nothing should prevent a proper compilation also under the Mac OsX operative system.

The third and last layer is called *High Abstraction Layer*. The HAL exposes the interface that all programmers should use to create specialised programs starting with the MAF. The general behaviour of MAF application is coded in class called *Logic* (Fig. 2). Logic communicates with four managers, VMEManager, ViewManager, OpManager, and InterfaceManager, which provide hardwired behaviour to all elements of the same type. To ensure the highest level of modularity, each manager is totally independent from the others and communicates only with its elements, respectively Virtual Medical Entities, Views, Operations and Interface Elements.

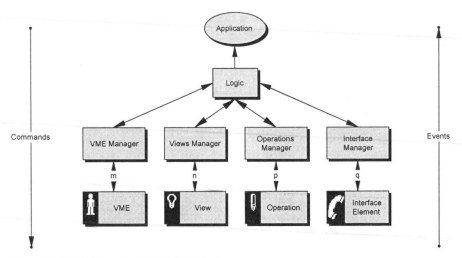

Figure 2. Logical structure of a MAF Application.

The communication takes place following a command-event model that ensures the child modules do not have to know their parent module in the architecture.

3.2 How to use the MAF

An application programmer can create a vertical application in two ways. The simplest is to create a new instance of Logic, define which Views and Operations it should make available and compile it. A new MAF application, fully based on the default MAF logic, is thus created with the following few lines of code:

```
// --- MyApp.h ---
#include wxWindows.h
#include mafServices.h

class MyApp : public wxApp
{
void OnInit();
void OnExit();
mafLogic *m_Logic;
}
```

Plus a few other lines of code to define, in the initialisation method, which Views and Operations the application should include:

```
// --- MyApp.cpp ---
#include Myapp.h

void MyApp::OnInit()
{
m_Logic = new mafLogic();
m_Logic->AddView(   new mafV1() );
m_Logic->AddView(   new mafV2() );
m_Logic->AddOperation(new mafOP1() );
m_Logic->AddOperation(new mafOP2() );
m_Logic->Show();
}
void MyApp::OnExit()
{
delete m_Logic;
}
```

After compilation the application programmer obtains a fully featured application, with all MAF services available and with only those views and operations that are relevant for the specific application.

The second approach to the creation of a vertical involves some customisation. Two levels of customisation are explicitly supported: the addition of new views and new operations developed starting from available templates, and the modification of Logic to obtain an application's behaviour different from the default one. While a default MAF application can be created by anybody able to run a compiler, the creation of new operations and views requires C++ fluency, and the customisation of Logic requires also a deep knowledge of the MAF internals. On the other hand, using the MAF any programmer can get in the quickest possible way a specialised application, with the level of customisation required.

To provide some examples of use, we can foresee the following scenarios:

- Biomedical Research user: a biomedical researcher with no interest in programming, willing to use the MAF technology in research activities. Biomedical researchers will find particularly interesting a MAF application called *Data Manager*, a default MAF application that exposes all views and all operations currently available. While complex in its use, because of the huge array of options, Data Manager is very powerful and helps to understand the potentialities of the MAF.

- Clinical user: a medical professional with no programming skills, willing to use the MAF technology in the clinical practice. This user will run vertical applications developed using the MAF. The first application to be made available will be the Hip-Op 2.0, a CT-based pre-operative planning environment for total hip replacement. Other similar applications are expected to come soon.

- Clinical Bioengineer: bioengineers active in clinical contexts may develop vertical applications aimed to support clinicians in various activities, by simply adding selected views and operations and recompile the default logic.

- Medical Technology Researcher: many researchers are developing complex algorithms to process or to visualise biomedical data. If they code their algorithms as additional Operations or Views, they can test them within a complete application environment, rather than wasting time to reinvent the wheel every time.

- Biomedical Software Developer: programmers who are developing professional computer aided medicine applications will find a great advantage in using the MAF. By adjusting the application logic to their need, and by adding new specialised views and operations, they can rapidly developing specialised applications focusing on the features that gives added value to their solutions.

- Multimod Developer: the MAF is distributed-ownership library. Thus, the most skilled programmers may join our Open Source program and see their bug fixes, improvements and additions included in the next MAF distribution, while retaining the ownership of the code they develop.

3.3   *Current Capabilities of the MAF*

The MAF is framework in its infancy, and we plan to expand it considerably in the next two years. In addition, once the MAF Open Source program is formally launched in late 2003 we hope to attract other programmers willing to contribute to the code base. Nevertheless, the MAF is already a very powerful tool that can be used to solve a variety of problems in the clinical practice and in the biomedical research. We shall list some of its capabilities and use the Data Manager to exemplify them.

An essential step in using the MAF is the ability to import into the VME Tree biomedical data. MAF imports 3D volumes (CT, MRI, PET, SPECT, 3D Ultrasounds) written in DICOM format, the international standard for medical imaging. Polygonal surfaces can be imported form STL or VRML files. STL format in particular can be used to exchange geometry with CAD programs, since MAF does not support mathematical surfaces such as NURBS that are commonly used in CAD programs. However, most STL exporters provided in commercial CAD are able to generate a tessellation of the surface with the required accuracy. On its side MAF can process efficiently very large numbers of polygons, so it is possible to import CAD models with a sufficient level of geometric accuracy. Standard 8 or 24-bit 2D images can be imported from JPEG, TIFF, GIF and BMP files. 16-bit diagnostic images can be imported only in DICOM format. Motion capture data can be imported as PGD and as C3D files. C3D files are translated into a sub-tree list of moving landmarks, which can be later clustered into a segmental motion using a specific operation. PGD format, less popular but more sophisticated, already provides the segmental trajectory and is translated by the MAF into a moving cluster of rigidly linked landmarks for each segment in the file. Any other dataset even if its not supported by the MAF, can be imported and is stored unchanged alongside with the other data. These unknown data are usually processed with an operation called *Open With External Program*, where the correct application is launched on the basis of the file extension. Thus, we can add to our database also an excel file containing all our calculation, which will be stored together with the other data into a single repository.

The efficacy of creating databases of disparate data is drastically extended by the ability of organising these VMEs into a hierarchical tree, the VME Tree. The pose matrix of each VME is interpreted as the pose of the VME with respect to the parent VME. In this way it is possible to create very complex spatial chains of datasets, one referred to the other. With the notable exception of motion data, imported datasets do not contain information on the pose of the dataset with respect to a global reference system. Thus, the relative VME is created attached to the root of the VME Tree, and an identity pose matrix is assigned to it. Afterward, I can cut and past or reparent the VME under another VME. Cut&Paste leaves the pose matrix unchanged, while reparent leaves the VME pose unchanged. This way I can keep the relative pose of the VME unchanged by attach it to another parent, or relocating the VME in the VME Tree without changing its absolute pose in space. Specific operations can modify the pose of a VME of known quantities or move it to a new pose in which the VME is aligned with another VME of the same type.

The entire VME Tree can be saved in the Multimod Storage Format. Currently, the directory structure of the MSF is based on the file system. Thus, saving a VME Tree creates a directory, not a file. Inside this directory there is an XML file that defines the VME Tree structure and all metadata of the VMEs forming it, and various sub-directories containing the raw data of each VME. By copying that directory and its content to another location we move the entire VME tree. We propose that zipped MSF files could become the *de facto* standard for exchanging biomedical data. This also considering that with the Data Manager, the data contained in a MSF file can be opened, visualised, but also exported into various standard formats.

The MAF provides very powerful ways to visualise these disparate data. Most views are able to render data of different types simultaneously, can display time-varying data, and support user interaction by pan, zoom and rotate of the camera. Surfaces can be displayed using a perspective surface view, which support multiple lights, fly-to camera motion and a multi-resolution scheme that ensure interactivity even with huge number of polygons. Volumes can be visualised with volume rendering views. The latest ray-casting view displays in the same view a CT or a MRI volume, together with polygonal surfaces. The volume render transfer function is very sophisticated and can make visible subtle differences, i.e. to visualise muscles in a CT Dataset. Another volume view is the isosurface view. This classic visualisation method, which use the Marching Cube algorithm to compute the polygonal surface that lies at a given density value in the volume, has been re-invented by introducing a new contour rendering technique. Contrary to the Marching cube algorithm, which required some minutes to run of full-resolution CT datasets, with this method the isosurface view becomes interactive and the user can adjust the density threshold interactively while observing in the view the result. The resulting isosurface can also be saved, providing also a first segmentation tool. A third volume visualisation method re-implements the so-called Digitally Reconstructed Radiograph (DRR) again to achieve interactive refresh rate with full resolution CT volumes. In this view the user can see the CT volume as a radiographic image projected from a point that can be changed interactively. The list is completed by simple views to display 2D images or single volume slices, or complex composite views that combine various modalities into the so-called *Multimodal Display* representations.

The Operations are the tools to be used to modify the VMEs. There are operations to edit the VME Data, to transform the entire dataset, to register the pose of two VMEs, etc. The Transform operation applies a generic affine transformation to the selected VME, defined with respect to any existent implicit or auxiliary reference system. After the transformation matrix is computed, a polar decomposition separates the roto-translation and the scaling components. The first pre-multiplies the VME pose matrix, defining the new pose; the second is simply trashed if the transformation was intended to be rigid (in this case the scaling is generated by round-off errors), or applied directly to the VME data if scaling was intended. The transformation can be defined not only in terms of matrix components, but also expressing rotations in terms of cardanic angles, Euler angles, attitude vector, quaternion, or helical axis. There are operations that create new VMEs, such as the virtual palpation operation, that places new landmark objects onto another VME. The virtual palpation is used for a variety of point-based anatomical registrations. It is possible to register onto a patient-specific skeletal anatomy an atlas of muscles insertions based on the Terry's musculo-skeletal database [5]. Motion data can be used to animate patient-specific bone models, by registering the landmarks palpated in the motion capture session with those virtually palpated on the bone models. Last, but not least, with virtual palpation the user can define anatomy-based auxiliary reference systems, and use them to define spatial transformations.

3.4   Future Developments

At the same time, the Multimod Consortium is working full time to continue the development of the MAF. While some future developments will be defined on the basis of the outcomes of current research activities, some of the major additions we plan to make can be anticipated.

A fundamental aspect of biomedical research that has been neglected in the MAF so far is simulation. While the MAF will never become a numerical simulation environment, it could evolve into a specialised environment for pre-processing medical data to form the simulation models, and to post-process the simulation results into alongside with diagnostic data, providing anatomically rich and clinically relevant visualisations of the simulation results. New VME types will be developed to support the finite element entities (nodes, elements, meshes, materials, and boundary conditions). Importers will be developed for finite element entities and for the associate simulation results. Specific views will visualise the results of the simulation in non-conventional ways. Support for multi-body dynamics solvers is planned in terms of import-export of geometry and of motion data. An essential element that is missing is a good mesh generation library. While the Multimod consortium does not plan to develop it, we shall consider the opportunity to integrate a third-party library.

On the visualisation front, some of the consortium members will work on the visualisation of deformable anatomical structures, such as muscles, so to ensure real-time interactive visualisation of deforming objects, even of their volume data. A possible objective is the creation of a digitally Reconstructed Fluoroscopy view, where a static CT dataset, a bones segmentation, and some motion data may be combined to create a synthetic fluoroscopy to show the joint motion.

A third development front comes from a distinct but related research project, called Multisense. The Multisense consortium is adopting the MAF to its scopes, and plans to extend it so to support multisensorial interfaces and virtual reality input-output devices, such as haptic or 3D trackers. This requires an architectural revision that is already in progress.

On the algorithmic front, we plan to develop a 2D-3D-registration algorithm, based on the very efficient DRR, which could be used to provide marker-less RSA solutions. A pre-operative CT scan of the joint would be used to create accurate models of the 3D skeletal anatomy, which would then be registered to post-operative 2D radiographs. The same operation would be repeated with the CAD models of the prosthetic components, which would be registered to their 2D image in the radiograph. Repetitions on follow-up radiographs would indicate if there is a change in the bone-implant relative pose, and thus implant migration.

4   The Living Human Digital Library

4.1   The Scope of the LHDL Project

The Living Human Digital Library aims to make Europe the world leader in the technologies for creating, maintaining and providing high-bandwidth and highly interactive access to large collections of coarse-grained, distributed, non-textual, multidimensional, time-varying resources, as required in a digital library for virtual laboratories. This general scope translates into a list of scientific and technological objectives:

- The creation of a digital library specifically designed to support a virtual laboratory, with all its aspects and needs. The technologies currently available can address only some of the requirements listed above and without any chance to integrate them into a single architecture.

- The provision of an array of value-added services that ensure an ever-expanding collection of scientific resources of universal relevance by motivating researchers to share their resources through the LHDL. All existing repositories for sharing scientific resources pose severe access barriers to most scientists, due to cumbersome interfaces, the limited control that the resource owner has on the sharing policy, strong centralisation, the lack of associated services such as collaborative work tools, etc. Because of these barriers only a few highly motivated researchers are sharing their data.

- The creation of all the technologies, services and policies that can ensure self-sustainability for the digital library after the end of the project. These technologies are commonly found in commercial services, such as those for selling 3D models for computer graphics, where the complexity and the heterogeneity of the resources and of the policies are much smaller. Scientists are sensitive to any restriction that may limit their autonomy, so the LHDL technology must provide high flexibility in its sharing policy and in the nature of the shared resources, while ensuring the accountability necessary to collect operational fees.

- The creation of all the technologies to allow the members of the virtual laboratory to provide added value services to clinical, industrial, cultural and societal customers. Most technology for digital collections currently available is limited to direct access to the resources. The LHDL aims to provide the resource owners with the technological support necessary to combine multiple resources, specific knowledge, RTD services and effective presentations into these added-value services.

- An architectural design sufficiently general and modular to ensure a total re-use of the LHDL technology in other application domains, and for industrial applications that may require the replacement of certain modules with legacy technologies.

4.2   Proposed Approach

These objectives can be pursued using four core technologies, at their maximum potential and with the highest level of integration: GRID, Semantic Web, Web Services and Remote Interactive Visualisation. GRID technology is required to provide high-bandwidth to large collections of coarse-grained, distributed, non-textual, multidimensional, time-varying resources. Semantic Web technology is required to add machine-understandable reasoning such as semantic search and sense making. Web Services technology is required to cope with the dynamic aspects of a digital library that provide as content, not only data, but also simulation services, collaborative work services, interactive visualisation services, etc. Remote Interactive visualisation is mandatory to inspect non-textual, large-sized, distributed resources.

Each of these technologies constitutes, by itself, a research topic attracting great attention in these days. Also, the integration of these technologies is at the top of the international research agenda of major groups: in particular, GRID support for web services, semantic web services, and semantic GRID are the official "buzzwords" at the core technology clubs in these days. However, the clear definition of the final objective ensures that a certain abstractness that affect these research fields will be avoided in the LHDL project. Early in the project, the consortium plans to make available:

- An environment for community building and collaborative work
- Four of the most valuable biomechanics data collections in the world
- Instruments to capture the complex operational ontology of the library
- A state-of-the art visualisation framework for biomedical data.

Sustainability at the end of the bootstrap phase is well accounted. The library should behave as a marketplace, in which sharing and return are key mechanisms for participation. The existence of a moderate subscription fee will be complemented by mechanisms, like tokens, based on ability and willingness to share resources in the community. The Consortium will identify or constitute the organisation that will operate the library indefinitely. Total flexibility in setting the access policies will allow each resource owner to define whatever criterion, ranging from totally free access to a fully commercial profile towards profit-making organisations. The exploitation committee will explore, from the outset, the many other business opportunities that such an enormous collection of resources related to the human body may have in other domains. A controlled access and accounted use of resources will foster new business initiatives in many fields (from medical, pharmaceutical, industrial, chemical and the like, to learning, and even leisure) and will also account for the sustainability of the required infrastructure in the long term.

4.3   A Brief Description of the Final System

The content of the LHDL will be a collection of resources and associated knowledge. These resources will be mostly large blocks of binary data, describing non-textual, multidimensional and time-varying information, or web services providing data processing and simulation services (Fig. 3).

This content will be created, maintained, and accessed via an Internet Semantic Portal, through which users and administrators will access all services. The Portal infrastructure will provide a user authentication mechanism that will give each user a high security single log-on to all LHDL services. The portal will also be a key instrument to disseminate to researchers, policy makers and the public at large the results of the project and those of the Biomechanics European Laboratory (BEL), the Virtual Laboratory Community that will operate the LHDL infrastructure. All security and trust issues will be fully addressed.

The BEL Virtual Laboratory will be supported by a complete set of community building tools, i.e. easy-to-use instruments to create and manage forums, real-time polls, collaborative editing of documents, and e-publishing. All Virtual Laboratory tools will be connected to a powerful Directory service, which will allow authorised users to control the access of other users to particular forums, polls, collaborative documents, and e-publishing resources. Supporting encryption mechanisms both during transmission and for stored information will further enforce the privacy. This will ensure that the Virtual Laboratory services are not only applicable for public, non-sensitive issues, but rather for the entire range of community needs, including those that require privacy such as industrial research, patient-specific information, initiatives of competitive nature within the community, etc.

Another aspect related to the community of users is the long-term sustainability of the LHDL. To this purpose, an accounting infrastructure will be integrated into the LHDL at an early stage in the project. Initially, the infrastructure will be used to support an automatic credit-debit model in which each resource owner sets the number of credits required to access the resource, and each registered user will have only a limited initial credit, thus promoting a sharing culture within the community. If you share your data, every time they are accessed you get more credits that you can use to access somebody else's resources. In a second phase, the e-commerce will allow the resource owner to set different prices for different user profiles (i.e. free for all researchers from my own university, 10 Euro for all non-profit-making institutions, and 100 Euro for all other users). Institutions will be allowed to buy credit blocks for their employees, if they use more than they provide. By the end of

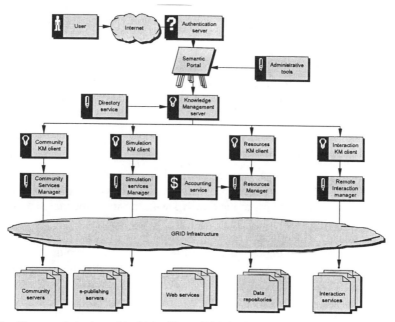

Figure 3. The general structure of the Living Human Digital Library

the project, we expect the turnover to be so high that a small support fee on all transactions should ensure the long-term self-sustainability of the LHDL. This long-term sustainability will be ensured not only by the income, but also by keeping the administration load as low as possible. To this end, a complete set of administrative tools will be developed and included to manage the Virtual Community services, accounting services, knowledge base, resource access services and resource management services.

A large collection of resources is useful only if each resource is linked to all of the relevant information. This is necessary not only to find what we want in the ocean of resources a library as large as the LHDL will contain, but also to make an effective use of that resource by having all related knowledge at hand. For this reason, a complete Knowledge Management infrastructure will be integrated into the LHDL with two precise objectives. A complete solution to create and maintain ontologies will ensure that the community itself sets the logic to be used in organising the LHDL resources. Such ontologies will allow the creation of a Semantic Web interface to the LHDL resources that the users will access to search for specific resources in an effective and meaningful way. Knowledge management tools will be developed and integrated to make it as simple as possible for the resource owner, or the rest of the community, to associate as much information as possible with the resource. Part of the metadata will be automatically generated by the system in relation to the resource attributes visible at the system level, or by processing eventual meta-information present in the resource. In addition, the owner will be able to link more knowledge to the resource, such as pertinent reports and papers, selected information from the patient health record, etc.

Quality Assurance will be addressed by the community itself, which will be allowed to associate with each resource further knowledge on how to use it, what is wrong with it, and so on. By combining knowledge management tools with tools from the Virtual Laboratory, it will be possible to enforce complex quality policies easily, as agreed by the community. This will be the main instrument to address one of the key concerns expressed by the whole biomedical community to the idea of sharing research data.

Once selected from the library, a certain resource will be inspected by the user, downloaded by the user, or combined with other resources to create a simulation. These three access modalities are associated with the three distinct sections of the LHDL architecture.

The Simulation Tools will ensure that three types of resources are effectively exchanged between the digital library and a simulation environment, at this stage represented by the actual patrimony of applications available within the community, which will be progressively included as binary objects in the library. By mean of the Simulation tools, the LHDL will be provided with all the services and interfaces to inter-operate with such a simulation environment. The LHDL and the simulation applications will exchange three types of resources: input data – data the application retrieves from the LHDL; simulation services – applications retrieved from the LHDL and executed; and simulation results – data returned from the application and stored in the LHDL.

Another way of accessing the selected resource is inspection. This means, in most cases, visualising the resource interactively. The Interactive visualisation tools will support remote visualisation and interaction, using a paradigm general enough to allow the support also of stereoscopic displays, 3D pointers, haptics, speech interactors, and other interaction modalities related to the augmented reality that may appear during the project.

The last access modality to an LHDL Resource will be the download. Here, the challenge will be the ability to ensure to the user the highest possible throughput, independently from its physical location. The Data Management Tools will constitute the backbone of the LHDL, and their realisation will require a significant part of the project resources.

The unique rationale between all these parts will be a GRID infrastructure, so all the services (community, knowledge management and accounting), tools (data management, interactive visualisation and simulation) and the portal itself will rely on a common GRID layer compliant to the Open Grid Service Architecture.

5 Conclusions

In this chapter we presented a general collaborative initiative called Living Human Project. We also described two concrete research actions, the Multimod Application Framework and the Living Human Digital Library, which could start-up this global initiative.

The Multimod Application Framework is in a well-advanced stage of implementation. The Multimod Consortium plans to release the code as an Open Source project within 2003. We hope the MAF will soon become the collective software framework through which a large number of researchers exchange, data, technology and knowledge. If this happens, the MAF may constitute an essential component for the Living Human Project. By sharing the same software tools, researchers may find easier to collaborate and share also their research results.

However, it is evident that it is only with the creation of an infrastructure such as the Living Human Digital Library that this Living Human Project may become a reality. This creates a major barrier for the project. To realise the LHDL it is necessary a long list of technological skills and a huge economical investment, that no single research group or single country can finance and realise. In the proposal we submitted to the European Commission we forecasted a total cost of nearly15 Millions Euro over five years (10 of EC support plus 5 of co-granting). The total value of the resources the project would mobilise in terms of data collections, usage of large-scale infrastructures, and value of the results from previous researches the partner planned to bring into the project, was predicted to be over 40 Millions Euro. To run the project we composed a consortium of 14 of the best research institutions in Europe.

It is difficult to imagine how this enormous endeavour may be realised without a substantial financial support from a trans-national agency such as the European Commission. We shall continue to monitor the grant opportunities so to identify other chances to get this support. In the meanwhile, the biomechanics and bioengineering communities should start to discuss if there is a possibility to realise some of the LHDL objectives. Maybe not through a huge monolithic project, but rather by assembling small local initiatives, supported by national or regional grants and loosely co-ordinated at the trans-national level, possibly with ad hoc resources from the EC for this purpose. The mini-symposium on the Living Human Project hosted inside the International Congress on Computational Bioengineering will be the first opportunity to discuss this possibility.

Acknowledgements

The authorship of this contribute does not reflect faithfully the long list of colleagues who are contributing to the initiatives listed here. We want to mention some of them here. The BioNet project was the brainchild of Gordon Clapworthy and Serge Van Sint Jan, respectively at the University of Luton and at the Free

University of Brussels. These two partners and their colleagues (in particular Alexander Savenko and Veronique Feipel) are now actively working in the Multimod Consortium together with the researchers at CINECA (Cinzia Zannoni, Marco Petrone, Silvano Imboden and Paolo Quadrani), at VICON (Lasse Roren) and at ESI (Muriel Beaugonin, Christian Marca). The LHDL proposal was draft by a number of esteemed colleagues working in 14 different institutions. Special thanks to Sanzio Bassini (CINECA), John Dominguez (Open University) and Marco Meli (EDW International) for the commitment during the final review session. Thanks also to the of the European Society of Biomechanics, which endorsed this initiative. Last, but not least, we have to acknowledge the essential contribution of all the colleagues that actively participated to the discussion forums we promoted so far on the BIOMCH-L mailing list, on the BioNet discussion groups, and on the BEL forum. A complete list would be impossible, so we just want to mention the moderators of BIOMCH-L, Ton Vander Bogert and Krystyna Gielo-Perczak, as well as Ilse Jonkers and Raj Mootanah, who are serving together with Serge Van Sint Jan as Delegates for the BEL Community.

References

1. Schroeder W., Martin K. and Lorensen B., *The Visualization Toolkit An Object Oriented Approach To 3D Graphics*, (Prentice Hall, Upper Saddle River 1998).
2. The Insight Tool Kit project: http://www.itk.org
3. Hudson T., Lin M., Cohen J., Gottschalk S. and Manocha D. V-COLLIDE: Accelerated Collision Detection for VRML, In: *Proc. of the second symposium on Virtual reality modeling language*, (ACM Press, New York 1997).
4. WxWindows Open Source project: http://www.wxwindows.org
5. Kepple T.M., Sommer 3rd H.J., Lohmann Siegel K., Stanhope S.J., A three dimensional musculoskeletal database for the lower extremities, *Journal of Biomechanics* **31(1)** (1998) pp. 77-80.

USE OF EXPLICIT FE IN THE PRE-CLINICAL TESTING OF TOTAL KNEE REPLACEMENT

MARK TAYLOR[1], RACHEL TAN[1], JASON HALLORAN[2] AND PAUL RULLKOETTER[2]

[1] *Bioengineering Science Research Group, School of Engineering Sciences, University of Southampton, Highfield, Southampton, SO17 1BJ. UK.*

[2] *Computational Biomechanics Group, School of Engineering and Computer Science, University of Denver, Denver, CO, USA.*

1   Introduction

The short and long term behaviour of total knee joint replacement is dependent on obtaining the optimal stress distribution within the bone-implant construct. The stress distribution within the prosthetic components, the bone-implant interface and within the supporting bone are ultimately dependent on the kinematics of the replaced knee. In turn, the kinematics are dependent on a number of factors including the geometry of the implant, particularly of the articulating surfaces, the relative alignment of the components, both with respect to each other and with respect to the bone, and tensions of the surrounding soft tissues. Of particular interest is the potential for wear of the polyethylene insert. Fatigue damage is likely to result in pitting or delamination and gross failure of the polyethylene component, whereas abrasive/adhesive wear may lead to an adverse osteolytic reaction and potential for implant loosening. Both failure modes will be influenced by the magnitude of the polyethylene stresses and the overall kinematics of the joint. Retrieval studies have shown that, unlike hips, the wear of TKR is highly variable [1,2] and this is probably due to the diverse kinematics which occur *in vivo*.

Fatigue damage and abrasive/adhesive wear of total knee replacements is a multi-factorial problem and is dependent on the material properties (including the molecular weight, degree of cross-linking, processing, sterilisation technique and counterface roughness), patient related variables (activity levels, which influence kinematics and loading), surgery related variables (orientation of the components and ligament balancing) and design related parameters (retention/resection of the posterior cruciate ligament, fixed or mobile bearing, degree of congruency). The relative contributions of these various factors to TKR wear are difficult to assess.

Currently, TKR designs are only subjected to a limited range of pre-clinical validation to assess their function, in terms of their kinematics and their potential long term wear performance. At the design stage, rigid body analyses may be performed to evaluate the kinematics of the design. At the prototype stage, a design will undergo wear testing. A number of experimental knee wear simulators have been proposed [3,4] with the aim of replicating the *in vivo* mechanical environment. The most complex simulators are capable of applying loads and motions in four principal directions (flexion-extension; inferior-superior and anterior-posterior translation and internal-external rotation). At present, TKR designs are evaluated under idealized conditions, typically simulating the bi-condylar loading of an average weight patient during level gait.

Knee simulators are important in pre-clinical assessment of the initial kinematics and wear of both new designs and materials. They have been used to examine the kinematics of different implant designs [5] and the long-term wear behaviour [3,6,7]. Although this gives a necessary benchmark for comparing different designs, it gives no indication of how a prosthesis is likely to perform in the diverse conditions likely to be experienced *in vivo*.

In addition to experimental studies, computer simulations can provide useful information to aid in the design process by evaluating potential TKR performance for a wider range of implant geometry and loading/kinematic conditions than can be examined experimentally. Until recently, simulations to predict implant kinematics and the stress distribution within the polyethylene insert were performed separately. However, as discussed earlier, the polyethylene stresses are intimately related to the kinematics of the joint. Early attempts to predict the polyethylene stresses were based on static FE analyses, which simulated the peak forces during the gait cycle. In these studies, the position of the femoral component relative to the tibia at that particular instant of the gait cycle had to be estimated and was assumed to be either in the deepest part of the polyethylene insert [8,9] or by estimating where contact would occur in a well function natural knee [10]. In an attempt to develop more realistic models, the relative positions of the prosthetic components at various stages of the gait cycle was derived from either rigid body kinematic analyses [11] or from fluoroscopy data [12]. For example, Sathasivam *et al.* [13] developed a three dimensional rigid body model with that reproduced the motions of the experimental Stanmore knee simulator [14]. The predicted motions then were used to define the relative positions of the prosthetic components within a three dimensional FE model and a series of static FE analyses were performed at various flexion angles to calculate the contact pressures and estimate sub-surface damage [11].

Only Estupinan *et al.* [15] and Reeves *et al.* [16] have modelled dynamic loading. Estupinan *et al.* [15] used an idcalized, two-dimensional model of a non-conforming knee replacement to simulate the influence of cyclic loading on polyethylene stresses. A 200 N load was applied to the femoral indentor which was then displaced 4 mm across the polyethylene surface, the load was removed and the indentor returned to its original position. Reeves *et al.* used a two dimensional sagittal plane model to examine the development of plastic strain in the polyethylene due to repetitive loading. The anterior-posterior motion of the femoral component was controlled by applying a displacement history from the literature.

The major limitation of all of these analyses are that they either assume the replaced knee to be statically loaded or in the rare cases where motion has been included, the motions are imposed by controlled displacements. *In vivo*, the motions of the knee are controlled by imposed forces, either directly by the joint reaction force or indirectly through the restraints of the surrounding ligaments.

At the University of Southampton and the University of Denver, we have been exploring the use of explicit FE analysis to study the performance of total knee replacement. Explicit finite element analysis has been specifically developed to simulate multi-contact and highly non-linear problems. It has been used extensively in the automotive industry to simulate impacts and to aid the development of occupant safety systems. The main advantage of explicit finite element analysis is that it is capable of calculating the kinematics and the implant stresses from a single analysis at a relatively low computational cost. The work described here will focus on the large body of work performed to date between the two laboratories on a particular TKR design (PFC Sigma, DePuy).

2   Simulation of a Gait Cycle in a Knee Wear Simulator

The initial explicit FE studies [17-21] were aimed at replicating the mechanical environment within the Stanmore knee simulator in order simulate a gait cycle [14]. The femoral component was modelled as rigid body using four-noded shell elements. The polyethylene was modelled as a deformable continuum assuming elastic-plastic material behaviour and meshed using continuum hexahedral elements (Fig. 1). The boundary conditions applied to the model were aimed at reproducing the mechanical environment existing in the Stanmore knee simulator [14]. The femoral component was allowed to translate in the inferior-superior direction, to rotate about a frontal axis to simulate valgus - varus rotation and to rotate about a tranvserse axis to simulate flexion-extension. The tibial insert was allowed to translate in the anterior-posterior direction and rotate about a fixed vertical axis located in the middle of the tibial condyles to simulate internal-external rotation.

Figure 1. Explicit FE model of tibiofemoral articulation.

In the experimental wear simulator, the horizontal component of the soft tissue restraints are represented by four springs, two anteriorly and two posteriorly. These springs were replicated within the FE model and assigned a stiffness of 10.4 Nmm⁻¹, and a corresponding rotational constraint of 0.3 Nm-deg⁻¹. The boundary condition loading histories for the axial force, internal-external torque, anterior-posterior force and the flexion-extension angle were defined according to the experimental protocol of the Stanmore knee simulator. The finite element simulations are typically performed at between 0.5 and 1 Hz. The models are then used to evaluate the resulting kinematics (anterior-posterior displacement, internal-external rotations), the contact pressures and areas, the polyethylene stresses and plastic strains. Analyses have been performed using either Pam-Crash-Safe or Abaqus explicit.

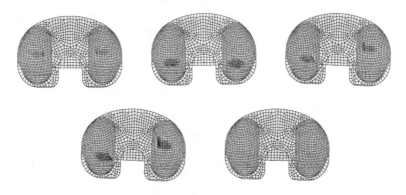

Figure 2. Contact pressure distribution at 5, 17, 35,45 and 65 % of the gait cycle (0 % heel strike, 60 % toe-off and 100 % heel-strike). Light gray corresponds to 0 MPa and black to pressures in excess of 20 MPa

The predicted contact pressure distribution on the proximal surface of the polyethylene insert from a typical analysis are shown in figure 2. At the beginning of the simulation, the medial and lateral condyles are slightly anterior of the neutral position. In the first 20 % of the gait cycle the femoral condyles moved posteriorly on the polyethylene component and the peak contact pressures rose to approximately 15 MPa. Between 20 % and 45 % of the gait cycle, the lateral femoral condyle moved anteriorly. During this phase, the peak contact pressure decreased to 10 MPa at approximately 30% of the gait cycle and then rose again to the maximum of 22 MPa at 45 % of the gait cycle. Between 45 % and 60 % the lateral condyle returned to a posterior position on the tibial component and the peak contact pressures decreased to 10 MPa. Both condyles remained in a slightly posterior position during the swing phase of gait (60 %-100 %) and the peak contact pressures remained constant at 5-6 MPa. In general, the predicted motions are similar to those seen experimentally, both in terms of general trends and the absolute magnitudes [17]. The influence of mesh density and contact algorithm on the predicted kinematics and contact stresses has been studied [17]. The kinematics were shown to be relatively insensitive to changes in mesh density, but the predicted peak contact pressure and contact area were found to be mesh dependent. To date, this technique has been used to examine the performance of TKR as a result of a range of parameters including eccentric loading [21], patient specific loading and tibial slope angle.

3   Influence of Unicondylar Loading on the Kinematics and Stresses Generated in a TKR

Normal alignment of the lower extremity is approximately 0 degree of mechanical alignment or 7 to 9 degrees of tibiofemoral anatomic valgus. The main objectives of total knee replacement (TKR) surgery are to alleviate pain and restore function by lower extremity re-alignment and soft-tissue balance. However, perfect knee alignment is not always obtained in total knee replacement. Varus-valgus malalignment has been shown to cause high contact stresses within the polyethylene insert [22] and in some cases may cause severe damage to the component.

The effects of eccentric loading, as a result of varus malalignment, were simulated by displacing the point of application of the axial load medially, along the flexion and extension axis. Two loading cases were simulated; loading offsets of 0 and 20 mm that resulted in medial:lateral loading ratios of 50:50 and 95:5 respectively, i.e bi-condylar and unicondylar loading, respectively. Because the femoral component is represented as a rigid body, offsetting the point of application of the vertical load meant that the center for varus and valgus rotations also was offset by the same amount.

In the knee joint, the collateral ligaments are positioned in the vertical direction to resist varus-valgus rotations.   Therefore, two models were considered: the standard horizontal linear spring model (HLS model), which replicates the mechanical environment of a knee wear simulator, and a model in which the springs were replaced by representations of the collateral ligaments (CL model). The collaterals were modelled as membrane elements, which allow buckling under compressive loads. The MCL and LCL had stiffnesses of 130 and 110 N/mm, respectively [23]. The proximal ends of both ligaments were rigidly attached to the femoral component and the distal ends were attached to the tibial component. The origins of the collateral ligaments were assumed to be approximately parallel to the vertical loading axis. Again, the predicted AP translations and the IE rotations of the prosthetic components were compared for the different models.

For the bi-condylar load cases, there was no significant difference in the kinematics of the TKR during the stance phase of gait between the HLS and CL models, although there were small differences during the swing phase (Fig. 3). For the unicondylar load case, there was a significant difference in the kinematics (Fig. 3). For both models, at the beginning of the simulation to 15 % of the gait cycle, the medial condyle moved from being centrally located to a slightly posterior position and the lateral condyle translated from the anterior edge to the posterior edge of the tibial component. As the simulation progressed to 45 % of the gait cycle, the position of the medial condyle for the two models still remained constant. However,

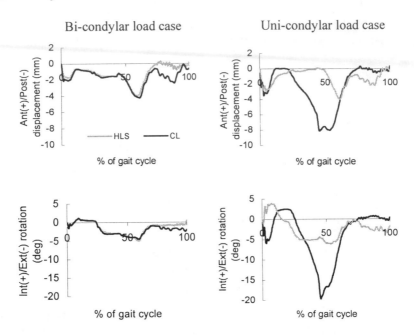

Figure 3. Predicted kinematics of the HLS and CL models when subjected to bi-condylar and uni-condylar load cases. Top row corresponds to the AP translations and the bottom row the IE rotations.

the HLS model internally rotated more than the CL model. The lateral condyle of the HLS model rode up the anterior lip of the tibial insert, nearly subluxing, between 45 % and 50 % of the gait cycle. The lateral condyle of the HLS model then moved back towards the center position at about 60 % of the gait cycle. During the swing phase, both condyles for this model remained at a slight posterior position. For the CL model, a similar pattern of motion was observed for the lateral condyle, but the degree of internal rotation was significantly less, with peak internal rotations of 18 and 7 degrees respectively for the HLS and CL models. The presence of the collateral ligaments constrains the vertical motion of the femoral component, reducing the amount the lateral condyle of the femoral component is able to ride up the anterior lip of the polyethylene, and hence limits the degree of internal rotation.

The maximum contact stresses for the HLS model were generally less than the CL model for the bi-condylar load case. The peak contact stress for the HLS model was approximately 22 MPa and occurred at 47 % of the gait cycle, as compared to the peak contact stress for the CL model of 27 MPa. There were only minor variations in the maximum contact stresses for the CL model between the unicondylar and bi-condylar load cases. However, for the HLS model, there was a significant increase between the bi-condylar and unicondylar load cases. The peak contact stress was 42.4 MPa at for the unicondylar load case and this was 12.4 MPa higher than the CL model.

The simple HLS model has been shown to be adequate for evaluating the ISO standard bi-condylar loading configuration. However, when considering abnormal loading, like uni-condylar loads, it is important to include representations of the collateral ligaments, so as to include the vertical constraint that they provide. If these are not included, such models tend to over-predict the amount of internal rotation and consequently over-predict the peak contact pressures that occur.

4 Influence of Patient Specific Loading

Although it is well understood that a TKR design will experience a range of loading conditions, at present they are evaluated under a limited range of idealised loading conditions. In order to assess the influence of variability in the kinematics and polyethylene stress a series of analyses was performed using subject-specific load cases. The knee joint reaction forces during walking for seven healthy subjects were calculated using an inverse dynamics method (Dr Patrick Costigan, Queen's University, Canada – Personal communication). The axial and anterior-posterior forces, internal-external torques and flexion angles calculated for these seven subjects are shown in figure 4, and these were used to drive the knee replacement model. The peak axial force was typically in the range of 1000-2000 N, although the predicted peak axial force for one individual was 2800 N. The anterior-posterior forces ranged from a posteriorly directed force of 250 N to an anteriorly directed force of 200 N. A consistent pattern of torque was observed, with the peak value for each individual ranging from 2 to 8 Nm. From figure 4, it can be seen that not all the subjects started at 0° of flexion. Two subjects started with a slightly hyperextended knee at approximately 2° and 5°, respectively. The other subjects began between 4° and 8.5° of flexion. Again, the anterior-posterior displacement, internal-external rotations and contact stress distributions of the knee prosthesis were reported for each individual.

In general, each individual's model produced a similar trend in both the anterior-posterior translations and the internal rotations. During the stance phase, the femoral component tended to translate posteriorly and internally rotate (Fig. 5). This was followed by an anterior translation and an external rotation during swing phase. Averaging this data and plotting the mean and standard deviation (Fig. 5) showed relatively minor variations between the kinematics of each individual. During the stance phase of the gait cycle, the standard deviation is on the order of 0.5 to 1 mm for the AP translations and 0.4 and 1 degree for the IE rotations.

Although the kinematics appear to be fairly consistent for this design for level gait, there was a larger range in the predicted maximum contact stresses (Fig. 5). The trend of the peak contact pressure during the gait cycle was fairly consistent, however, the magnitude was sensitive to the magnitude of the axial force during the

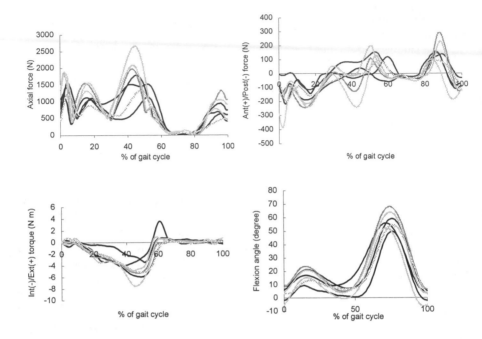

Figure 4. Force data for seven healthy individuals applied to the TKR model (Dr P. Costigan, Queen's University, Canada – Personal communication).

first 30 % of gait cycle the standard deviation was between 2 and 5 MPa and between 30 and 60 % of the gait cycle it varied from 1 to 4 MPa. For this particular design it has been shown that during level gait the kinematics are relatively insensitive to variation between patients, whereas substantial variations may occur in the predicted peak contact pressure. Other designs may be more or less sensitive to patient related variability. It may also be true that there will be greater variability between subjects for different activities, for example stair ascent. By using patient specific forces, it is possible to assess the performance envelope for a total knee replacement design.

5   Influence of Posterior Tibial Slope

Posterior tilting of the tibial tray and insert is frequently used clinically to facilitate a greater range of flexion by promoting natural femoral rollback, and to resist anterior tibial subsidence by optimizing the underlying bone strength following resection. Significant posterior tilt angles may also be included unintentionally. As a result of the altered tibiofemoral constraint and conformity, relative motion and

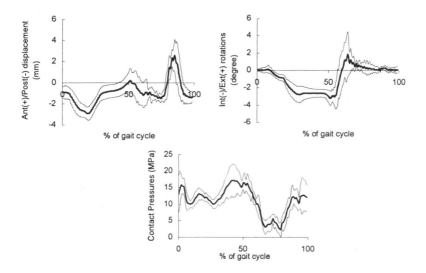

Figure 5. Variation in the predicted kinematics and the peak contact pressures as a result of patient specific loading. The thick line represents the mean for the 7 patients and the thin line represents ± one standard deviation of the mean.

insert stresses will also change. Since, as mentioned previously, the joint kinematics and contact mechanics will play a large role in the long-term success of total knee arthroplasty, it is important to quantify the effect of posterior tilting on these measures. Thus, the effect of posterior tilt on joint mechanics during *in vitro* physiological loading and the sensitivity of various current TKR devices to these changes were evaluated.

The TKR was analyzed during gait simulation at successive, even posterior tilt angles from 0° (no posterior slope) until component dislocation occurred during the gait cycle. In each analysis, the femoral component was loaded at full extension to find a stable initial position before applying the simulated gait loading cycle. During the initial loading cycle, a nominal load was applied to the femoral component while it was unconstrained in the AP and varus-valgus degrees of freedom. This initial loading resulted in settling of the femoral component into the lowest point in the tibial dish. For every tilt angle the simulated soft-tissue constraint remained in the original, 0° plane in order to evaluate changes in joint mechanics as a result of the changing posterior tilt. Again, tibiofemoral AP displacement, IE rotation, insert stresses, and contact area were recorded as a function of the gait cycle for each analysis.

Changes in the posterior tibial slope varied tibiofemoral constraint and hence, relative kinematics. Both changes in relative kinematics and tibiofemoral conformity were factors in increasing surface and subsurface insert stresses. The semi-constrained PCR design showed only small changes in AP motion from 0° to 10° tilt, generally less than 0.5 mm (Fig. 6). From 12° to 16° tilt the peak anterior tibial motion increased by 3.1 mm due to the decrease in anterior constraint. The total range of AP motion was very consistent throughout the range of tilt angles studied (Fig. 6). Femoral component subluxation occurred posteriorly during the gait cycle at 18° tilt. Both the total range and peak internal rotation decreased steadily with increasing tilt. The peak internal rotation decreased from 5.2° with no posterior slope to 2.3° at 16° tilt (Fig. 6). The total IE range of motion decreased nearly linearly from 6.3° to 2.7°, or approximately –0.22° IE/°tilt (Fig. 6). Peak contact pressures during the cycle were steady, then increased linearly after 10° tilt due to variation in kinematics creating posterior edge loading (Figs. 7, 8). Near 55% of the gait cycle the 0°, 2°, and 4° analyses experienced a spike in the contact pressure results due to relatively high IE rotation creating poor-conformity contact. The reduction in IE rotation with posterior tilt angle reduced this peak by approximately 25% for the 8° and 10° analyses (Fig. 7). Although there were only small changes in AP kinematics for the range of tilt studied, the composite von Mises stress distribution (all stress contours added for each increment of the gait cycle) shows a progressive posterior motion of the insert stresses with increasing posterior tilt (Fig. 8).

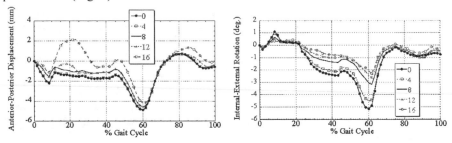

Figure 6. Influence of posterior slope on the predicted kinematics.

This particular design of TKR was found to be relatively accommodating to changes in tilt in the range that would be used clinically. Peak contact pressure remained steady until beyond 10° of posterior tilt. Predicted joint kinematics were fairly consistent, again, especially through 10° of tilt. The peak and range of IE rotation decreased consistently with tilt. Increased anterior insert motion marked the rise in contact pressure realized at 16° posterior tilt as a result of posterior edge contact. The single sagittal insert radius was relatively insensitive to changes in tilt that could occur when surgically reproducing the anatomic tibial slope in TKA.

Wasielewski and coauthors [24] found a statistically significant correlation between tibial insert posterior slope and increasingly posterior articular wear track location with an unconstrained insert. This is consistent with the composite von Mises and contact stress distributions.

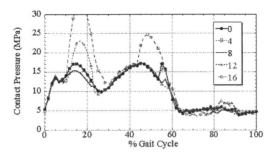

Figure 7. Influence of posterior slope of on the predicted peak contact pressures

The results from this study demonstrate that posterior tilting of the tibial tray and insert in the range used clinically may alter the tibiofemoral AP and IE constraint, and result in significant changes in stability and predicted kinematics during gait simulation. Corresponding contact mechanics may also be substantially altered due to variation in conformity, especially when edge-loading conditions appear. Additionally, the unconstrained PCR designs, which are frequently implanted with a posterior tilt, may be more sensitive to these changes.

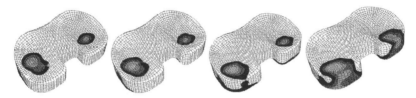

Figure 8. Composite von Mises stress distribution (summation of all the stress contours for each increment of the gait cycle)

6 Rigid body Analysis

Although the deformable, explicit FE models are reasonably efficient, reported CPU time is still in the range of 6 hours to more than a day for a full gait cycle [17,25]. Parametric analysis of component positioning, numerical wear simulation of TKR components, or other studies which require many repeated analyses or real-time feedback can therefore be cost prohibitive. Modelling the polyethylene insert as a rigid body, together with an elastic contact definition, has been reported to reduce the solution times to less than thirty minutes. However, potential differences in predicted kinematics between rigid and deformable insert representations are currently not well understood, and it is unclear if the estimates of contact pressure distribution are acceptable given the significant reduction in computational time. The explicit analysis permits the polyethylene insert to be easily characterized as a deformable or rigid body. In order to estimate the contact pressure distribution during a rigid body analysis, softened contact capability was employed. A relationship between contact pressure and surface overclosure (penetration) was estimated for the rigid-body contact definition based on the polyethylene material data. The "bed of springs" method from elastic foundation theory was used:

$$Pressure = \frac{(1-v)E}{(1+v)(1-2v)} \cdot \frac{overclosure}{thickness} \tag{1}$$

where E is the Young's modulus of the insert and v is Poisson's ratio. The initial slope of the elastic-plastic material model (571.6 MPa), Poisson's ratio (0.45), and average thickness of the insert (10 mm) were used to calculate the linear pressure-overclosure relationship.

The deformable material behavior is nonlinear, and therefore beyond the initial linear region the rigid-body pressures calculated using the linear pressure-overclosure relationship will be overestimated.

In order to most accurately perform parametric or repetitive studies using the rigid body approximations, such as numerical wear simulation with many identical repeated analyses, more precise predictions of contact pressures and areas are required. In an effort to enhance the results with the rigid-body simulations, the pressure-overclosure relationship was optimized by minimizing the difference between the contact pressure and area results determined using a rigid and deformable analysis. To accomplish this, a nonlinear constrained optimization routine in the IMSL library (Visual Numerics, Inc., San Ramon, CA) was used in conjunction with a C program to integrate IMSL and Abaqus. The objective function to be minimized calculated the sum of the percent difference between the rigid and deformable results for twenty points along the contact pressure and area results curves.

The rigid body and fully deformable analyses predicted very similar kinematics. The AP motion was nearly identical, and the IE rotation had a maximum difference of less than 0.5° (Fig. 9). Peak contact pressures were found near 15 %, 45 % and 55 % of the gait cycle, and were approximately 17 MPa for the deformable analysis (Fig. 10). The initial, 'bed-of-springs' estimate of the pressure-overclosure relationship resulted in very close agreement for the contact area and consistent overestimation of the peak pressures, although the general trends were acceptable (Fig. 10). Once the rigid body pressure-overclosure relationship had been optimized, there was close agreement with the fully deformable analysis results. The contact pressure contours for positions throughout the cycle were compared and closely match (Fig. 10). CPU time for the full deformable analysis was 6 hours, yet it was only 8 minutes for the rigid body analysis, nearly a 98 % reduction.

The pressure-overclosure model had little bearing on the predicted kinematics. Thus, if only kinematics are of interest in an analysis, an efficient rigid body model would provide acceptable results in this range of physiological loading conditions. The linear, 'bed-of-springs' contact pressure-surface overclosure relationship was a reasonable, first-order approximation to the deformable pressure distribution. Optimising the overclosure parameters resulted in excellent agreement between the deformable and rigid analyses. Limitations of the rigid body approach include significant overprediction of contact pressure with low conformity contact situations, such as the edge-loading conditions. Also, internal insert stresses and strains are not calculated in any of the rigid body analyses, nor is there currently the potential for including any time-dependent material properties, such as creep. In addition, the relationship was optimized for a particular mesh and loading conditions, and may change with significantly higher loads or variation in geometry or mesh density. However, given the vast reduction in computational time of approximately 98 %, the rigid body approach will be useful in a variety of situations, such as numerical wear simulation that repeatedly applies the same loading conditions, or investigating the effects of variable component positioning.

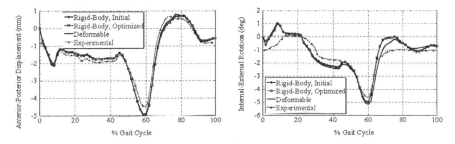

Figure 9. Comparison of the kinematics predicted by a rigid body model and a deformable model.

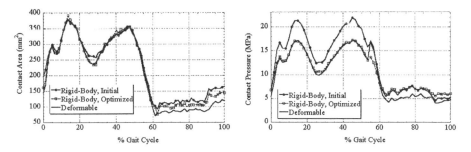

Figure 10. Comparison of the contact areas and pressures predicted by a rigid body model and a deformable model.

7 Numerical Wear Simulation

Mechanical wear simulators have been developed to this end, but require long testing times and are not ideal for investigating variables of implant geometry or effects of relative implantation locations. In addition, wear tests are slow and have a high associated cost. Tests are performed at a frequency of 1 hertz, which equates to 11.5 days/million loading cycles or 115 days for a typical 10 million cycle test. The average person with a joint replacement walks 1-2 million steps per year [26], so current tests equate to 5-10 years of normal activity. In order to simulate 20 years of service life, typical for a young TKR patient, these tests should be extended to at least 20-40 million cycles. This is clearly impractical with current knee wear simulators. Current test methods only simulate the gait cycle and recently stair descent [27]. In everyday life, the knee is subjected to a range of activities, including ascending stairs and rising from a chair. These activities, although less frequent, are more demanding than level gait, as they have higher flexion angles and higher loads. The experimental wear simulators may be considered to be semi-constrained. The flexion-extension and internal-external axes of rotation are fixed, where as the knee is free to decide the point about which it rotates. This artificial constraint may influence the wear process, particularly for higher flexion activities like ascending/descending stairs. Current experimental wear tests are performed under 'ideal' conditions, with respect to the implant orientation and the applied loading and boundary conditions (simulated soft-tissue constraints and axes of rotation). The prosthetic components are orientated within the machine in a neutral position. The loads are assumed to be for an average individual with a mass of approximately 70 kg and are typically distributed 50:50 to the medial and lateral condyles. The constraints representing the soft-tissue assume that the medial and lateral sides are balanced.

These ideal conditions are rarely achieved in vivo. Fluoroscopy studies have shown that lift-off often occurs, causing uni-condylar loading of either the medial or lateral compartment. Gait studies have shown that the gait patterns can vary significantly between individual strides within same patient and between patients, particularly in the early post-operative period. This will lead to variations in the ratios of the axial force to the anterior-posterior loads and internal-external torques. Patient mass can vary significantly, with masses in excess of 100 kg frequently occurring. In order to address these issues there is a need to develop numerical simulations to compliment experimental wear testing.

In an effort to produce results efficiently, numerical wear simulation methods have been developed based on Archard's law [28], and used in conjunction with implicit FE analyses [29]. The development of the explicit FE analysis has now made it practical to simulate the wear of knee joint replacements during force-controlled loading conditions. The explicit FE wear simulation can include important changes in relative kinematics due to changes in insert geometry during evaluation of long-term wear behaviour.

Although it is of future interest to perform long-term wear simulation, the initial rigid and deformable analyses discussed previously were used to estimate wear of the tibial inserts for 1 million loading cycles from a single gait cycle analysis. Predicted wear was enforced on a nodal basis, and linear penetration at each node was estimated using Archard's law [28]:

$$W = k \times s \times p \tag{2}$$

where W is the linear wear penetration, s is the tangential sliding distance, p is the contact stress, and k is the wear coefficient. k was adapted from Wang et al. [30] for the TKR analyses. Using a custom script, the sliding distance and contact stress data were extracted from the FE results and were used to estimate linear wear values for each node on the articular surface of the polyethylene component. Linear wear was applied to each node in a direction normal to the surface. The nodal positions were recorded before and after application of the simulated wear and the linear and volumetric wear were estimated.

Maximum linear penetration was determined for two designs, the semi-constrained PFC Sigma and an unconstrained TKR for a simulated 1 million gait cycles (Fig. 11). The unconstrained insert experienced greater contact pressures and smaller contact areas due to lesser tibiofemoral conformity. This resulted in larger predicted linear wear, but a 10 % decrease in wear volume when compared to the more conforming, semi-constrained device. For the semi-constrained insert, a 3.2 % difference was found in the peak penetration using rigid and deformable insert representations. For the unconstrained insert, the difference was only 2.2 %. Corresponding volumetric wear predictions varied by 6.2 % (semi-constrained) and 4.2 % (unconstrained). A significant reduction in CPU time was realized using the rigid body approach with optimized pressure-overclosure relationship, which completed in approximately 15 minutes compared to more than 6 hours for the deformable model.

The predicted wear trends of increasing volumetric wear with larger areas of contact is similar to results found on retrieved TJR components, and determined using mechanical simulators [31,32]. Work is currently in progress to develop adaptive FE simulations of wear, which will be able to account for the changes in geometry, and the corresponding changes in kinematics and contact pressures, that occur during extended wear testing (20+ million cycles). In addition, experimental validation by comparison with a knee wear simulator is ongoing, and will include both volumetric wear measurement (from weight loss) and surface profile measurement (includes linear wear and creep).

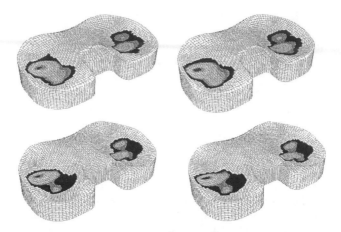

Figure 11. Wear contours determined using deformable (left column) and rigid body (right column) analyses for semi-constrained (top) and unconstrained (bottom) designs.

8   Full TKR Model: Analysis of the Patello-Femoral Kinematics and Stresses

Solid models of the femur, patella, tibia, and fibula were developed from computed tomography (CT) scan data, obtained from the Visible Human Female dataset. The bone models were imported, placed appropriately and trimmed using standard surgical technique, and meshed with rigid elements (Fig. 12). Membrane elements were chosen to represent the quadriceps tendon and ligamentous structures (Fig. 12). These structures include contact to capture the effects of soft-tissue wrapping on the implant and bone [23,33]. Bony protrusions of the solid models were used as insertion sites for the soft-tissue structures. Nonlinear elastic material definitions for the soft-tissue structures were also adopted from available literature [34,35]. Separate load-displacement analyses were used to validate the response of each ligament, including the quadriceps tendon, patellar ligament, lateral and medial collaterals, and posterior cruciate ligament.

In order to describe the relative kinematics for the TKR model in a manner consistent with experimental studies, a three-cylindric model of joint motion was incorporated for both tibio- and patello-femoral motion [36]. Coordinate frames were created with nodes placed on the femur, tibia, and patella, and relative kinematics were measured and described as three rotations and three translations. The model of joint motion was implemented with a post-processing script written in Python to interface with the output results database.

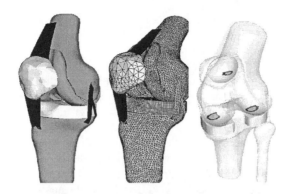

Figure 12. TKR model solid models (left), finite element mesh (middle), and transparent view showing patellar button and contact pressures (right).

Experimental kinematic data were obtained from articulation of the TKR device during gait simulation using the Purdue knee simulator [37]. Two implanted cadaveric knees were run under level walking conditions in the knee simulator. The Purdue simulator applies a vertical load and flexion angle at a simulated hip, and a tibial torque and a medial-lateral directed force at the representative ankle. A quadricep actuator balances the vertical load through the patellar ligament. Six-degree-of-freedom spatial linkages were mounted between the femur and the patella and femur and tibia. Provided experimental data, therefore, included tibio-femoral and patello-femoral kinematics and measured feedback of quadricep load as a percent of gait cycle. Hence, in order to verify model-predicted patellar motion the experimental tibio-femoral motion and quad load were used as inputs to the TKR model, and predicted patellar kinematics were compared to experimental data.

Boundary conditions were applied to reproduce the Purdue knee simulator environment. Measured tibial flexion-extension and internal-external rotations, anterior-posterior and medial-lateral translations, and vertical and quadriceps loading were applied to the model. In order to simulate the changing orientation, the applied quadriceps load is coupled to the flexion-extension angle as occurs in the experimental simulator. The Q-angle of the simulator was also included. The predicted patello-femoral kinematics were compared with the experimentally determined six degree-of-freedom kinematics using both a relatively coarse and a fine button mesh. A deformable as well as a rigid body analysis was performed using each mesh. Finally, contact area and contact pressure were compared between rigid body and deformable analyses.

Patello-femoral kinematic verification showed good agreement between experimental and model predicted results for both trends and magnitudes (Fig. 13). The plotted experimental data is the average of the two measured knees. Predicted patellar flexion matched experimental results to within approximately 1.3°. Flexion results from both meshes and both rigid and deformable analyses were very similar (Fig. 13). Model predicted and experimental patellar spin were both near 1°. The largest discrepancy between model and experimental data is seen in the medial-lateral tilt of the patella. The model over predicted medial tilt during flexion by up to 2.7° (Fig. 13). The two data sets were consistent in all degrees of freedom except tilt; one knee demonstrated a similar pattern to the model but the other exhibited almost no tilt. Predicted relative translations also correlated well with the experimental data, with peak differences of 1.7 mm along the femoral inferior-superior axis, and 2.7 mm along the femoral anterior-posterior direction. Predicted and experimental medial-lateral shift were both under 1 mm. Again, kinematic predictions resulting from the rigid body and deformable analyses were almost indistinguishable.

CPU time for the patello-femoral verification deformable analysis (coarse mesh) required approximately 18 hours and the corresponding rigid body representation required 2 hours.

Use of the explicit code allows 'buckling' of the soft tissues to occur without the stability problems that would take place with implicit analyses. During flexion activity this can act to simulate the recruitment of different ligament fibers [33]. This initial use of explicit FE modeling of soft-tissue structures is advantageous and will likely continue to progress. The addition of time-dependent material models for the soft tissues should be an area of future development.

In order to perform a rigorous experimental verification, the tibio-femoral and patello-femoral articulations were initially examined separately. Additionally, although the collaterals and posterior cruciate load-displacement behaviour was verified independently, the knee constraint provided was not included in the present experimental verification. The rigid bone models were used primarily for soft-tissue attachment sites. Futher development of the knee model will include a distribution of bone material properties, and a focused study on verification of soft-tissue constraint before thorough experimental verification of predicted simultaneous, force-controlled articulation.

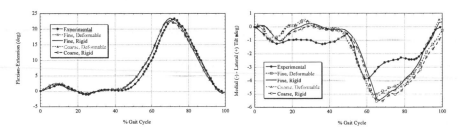

Figure 13. Experimental (Purdue knee simulator) and model predicted patello-femoral a) flexion-extension and b) medial-lateral tilt as a function of gait cycle.

9 Conclusions

Joint kinematics and contact mechanics substantially influence the short- and long-term performance of current total knee replacement devices. Although a multifactorial problem, implant contact stresses and relative motion play a significant role in wear and fatigue damage to the polyethylene insert, potentially limiting the useful life of the implant. Fatigue damage, such as pitting or delamination, can lead to catastrophic insert failure, while excessive abrasive/adhesive polyethylene wear may initiate an osteolytic reaction to particulate debris, contributing to eventual loosening and revision surgery. A volume of recent research has, therefore, focused on the prediction of kinematics at the joint interfaces and contact stresses and areas due to articulations as an indication of potential clinical performance.

Until recently, TKR stress analyses were generally performed using implicit FE analysis to predict surface and subsurface implant stresses during displacement-controlled conditions, and kinematic analyses were frequently formulated as rigid body dynamics problems to estimate relative motion during various loading conditions. Explicit FE models may be advantageous due to their stability during general, force-controlled analysis and ability to simultaneously predict implant stresses and relative motion. In addition, explicit FE formulations are unique in that they generally allow for either rigid or deformable representations of meshed components.

In our work, explicit FE formulations have been developed for both tibiofemoral and combined tibio- and patellofemoral articulations. Verification results from the computational TKR models showed that the predicted patellofemoral and tibiofemoral kinematics were in good agreement with experimental knee simulator measurements. Current studies have demonstrated the significance of eccentric loading and posterior tilting, and ongoing research includes the analysis of implant design parameters, sensitivity to implantation position, and effects of surgical procedures.

References

1. Engh G.A., Failure of the Polyethylene Bearing Surface of a Total Knee Replacement within 4 Years, *Journal of Bone and Joint Surgery-American* **70A** (1988) pp.1093-1096.
2. Blunn G.W., Joshi A.B., Lilley P.A., Engelbrecht E., Ryd L., Lidgren L., Hardinge K., Nieder E., Walker P.S., Polyethylene Wear in Unicondylar Knee Prostheses - 106 Retrieved Marmor, Pca, and St-Georg Tibial Components Compared, *Acta Orthopaedica Scandinavica* **63** (1992) pp. 247-255.
3. Burgess I.C., Kolar M., Cunningham J.L., Unsworth A., Development of a six station knee wear simulator and preliminary wear results, *Proceedings of the Institution of Mechanical Engineers*, **Part H** (1997) pp. 211.

4. Walker P.S., Blunn G.W., Perry J., Bell C.J., Sathasivam S., Andriacchi T.P., Paul J.P., Haider H., Campbell P.A., Methodolgy for long-term wear testing of total knee replacements, *Clinical Orthopaedics and Related Research* **372** (2000) pp. 290-301.
5. DesJardins J.D., Walker P.S., Haider H., Perry J., The use of a force-controlled dynamic knee simulator to quantify the mechanical performance of total knee replacement designs during functional activity, *Journal of Biomechanics* **33** (2000) pp. 1231-1242.
6. Barnett P.I., McEwen H.M.J., Auger D.D., Stone M.H., Ingham E., Fisher J., Investigation of wear of knee prostheses ina new displacement/force controlled simulator, *Proceedings of the Institution of Mechanical Engineers*, **Part H 216** (2002) pp. 51-61.
7. McGloughlin T., Kavanagh A., Wear of ultra-high molecular weight polyethylene (UHMWPE) in total knee prostheses: a review of key influences, *Proceedings of the Institution of Mechanical Engineers*, **Part H 214** (2000) pp. 349-359.
8. Bartel D.L., Bicknell V.L., Ithaca M.S., Wright T.M., The effect of conformity, thickness and material on stresses in ultra-high molecular weight components for total joint replacement, *Journal of Bone and Joint Surgery* **68A** (1986) pp. 1041-1051.
9. Mottershead J.E., Edwards P.D., Whelan M.P., English R.G., Finite element analysis of a total knee replacement by using Gauss point contact constraints, *Proceedings of the Institution of Mechanical Engineers*, **Part H 210** (1996) pp. 51-63.
10. Bartel D.L., Rawlinson J.J., Burstein A.H., Ranawat C.S., Flynn W.F., Stresses in polyethylene components of contemporary total knee replacements. *Clinical Orthopaedics and Related Research* **317** (1995) pp. 76-82.
11. Sathasivam S., Walker P.S., Computer model to predict subsurface damage in tibial inserts of total knees. *Journal of Orthopaedic Research* **16** (1998) pp. 564-571.
12. Ishikawa H., Fujiki H., Yasuda K., Contact analysis of ultrahigh molecular weight polyethylene articular plate in artificial knee joint during gait movement, *Journal of Biomechanical Engineering-Transactions of the Asme* **118** (1996) pp. 377-386.
13. Sathasivam S., Walker P.S., A computer model with surface friction for the prediction of total knee kinematics, *Journal of Biomechanics* **30** (1997) pp. 177-184.
14. Walker P.S., Blunn G.W., Broome D.R., Perry J., Watkins A., Sathasivam S., Dewar M.E., Paul J.P., A knee simulating machine for the performance evaluation of total knee joint replacements, *Journal of Biomechanics* **30** (1997) pp. 83-89.
15. Estupinan J.A., Bartel D.L., Wright T.M., Residual stresses in ultra-high molecular weight polyethylene loaded cyclically by a rigid moving indentor in

non-conforming geometries, *Journal of Orthopaedic Research* **16** (1998) pp. 80-88.

16. Reeves E.A., Barton D.C., FitzPatrick D.P., Fisher J., A two-dimensional model of cyclic strain accumulation in ultra-high molecular weight polyethylene knee replacements, *Proceedings of the Institution of Mechanical Engineers*, **Part H 212** (1998) pp. 189-198.

17. Godest A.C., Beaugonin M., Haug E., Taylor M., Gregson P.J., Simulation of a knee joint replacement during a gait cycle using explicit finite element analysis, *Journal of Biomechanics* **35** (2002) pp. 267-276.

18. Halloran J., Patrella A., Rullkoetter P., *Explicit finite element model predicts TKR mechanics*. in *49th ORS*. (2003).

19. Halloran J., Rullkoetter P., *A dynamic finite element model for prediction of TKR mechanics*. in *48th ORS*. (2002).

20. Halloran J., Rullkoetter P.J., *Development of an explicit finite element based total knee replacement model*. In *ASME Bioengineering Conference* (2001).

21. Taylor M., Barrett D.S., Explicit finite element simulation of eccentric loading in total knee replacement, *Clinical Orthopaedics and Related Research* **414** (2003) pp. 162-171.

22. Liau J.J., Cheng C.K., Huang C.H., Lo W.H., The effect of malalignment on stresses in polyethylene component of total knee prosthese - a finite element analysis, *Clinical Biomechanics* **17** (2002) pp. 140-146.

23. Mommersteeg T.J.A., Blankevoort L., Huiskes R., Kooloos J.G.M., Kauer J.M.G., Hendriks J.C.M., The Effect of Variable Relative Insertion Orientation of Human Knee Bone-Ligament-Bone Complexes on the Tensile Stiffness, *Journal of Biomechanics* **28** (1995) pp. 745-752.

24. Wasielewski R.C., Galante J.O., Leighty R.M., Natarajan R.N., Rosenberg A.G., Wear patterns on retrieved polyethylene tibial inserts and their relationship to technical considerations during total knee arthroplasty, *Clin Orthop* **299** (1994) pp. 31-43.

25. Giddings V.L., Kurtz S.M., Edidin AA, Total knee replacement polyethylene stresses during loading in a knee simulator, *Journal of Tribology* **123** (2001) pp. 842-847.

26. Wallbridge N., Dowson D., The walking activities of patients with artificial joints, *Eng. Med.* **11** (1982) pp. 96.

27. Benson L., DesJardins J.D., LaBerge M., *Stair descent loading pattern for TKR wear simulation*. In *48th Annual Meeting of theOrthopaedic Research Society*.Dallas,USA (2002).

28. Archard J.F., Contact and rubbing of flat surfaces, *Journal of Applied Physiology* (1953) pp.981-988.

29. Maxian T.A., Brown T.D., Pedersen D.R., Callaghan J.J., A sliding-distance-coupled finite element formulation for polyethylene wear in total hip arthroplasty, *Journal of Biomechanics* **29** (1996) pp. 687-692.

30. Wang A., Lubrication and Wear of Ultra-High Molecular Weight Polyethyelene Total Joint Replacements, *Tribology International* **31** (1998) pp. 17-33.
31. Gillis AM , and Li S, A Quantitative Comparison of Articular Surface Damage Areas and Types in Retrievals From 4 Different Knee Designs, *Trans 45th Annual Meeting ORS* **45** (1999) p, 861.
32. Harman M.K., Banks S.A., Hodge W.A., The Influence of Femoral Geometry on In-Vivo Kinematics and Wear in Two Designs of PCL-Retaining Total Knee Arthroplasty, *Trans 45th Annual Meeting* **45** (1999) pp.148.
33. Blankevoort L., Huiskes R., Ligament-bone interaction in a three-dimensional model of the knee, *J. Biomech. Eng.* **113** (1991b) pp. 263-269.
34. Staubli H.U., Schatzmann L., Brunner P., Rincon L., Nolte L.P., Mechanical Tensile Properties of the Quadriceps Tendon and Patellar Ligament in Young Adults, *American Journal of Sports Medicine* **27** (1999) pp. 27-34.
35. Quapp K.M., Weiss J.A., Material characterization of human medial collateral ligament, *J. Biomech. Eng.* **120** (1998) pp. 757-763.
36. Grood E.S., Suntay W.J., A Joint Coordinate System for the Clinical Description of Three-Di,mensional Motions: Application to the Knee, *Transactions of the Associated Society of Mechanical Engineers* **105** (1983) pp. 136-144.
37. Zachman N.J., Hillberry B.M., Kettlekamp, Design of a Load Simulator for the Dynamic Evaluation of Prosthetic Knee Joints, *ASME publication* 78-DET-59 (1978).

THE PLUMBING OF LONG BONES

STEPHEN C. COWIN

The New York Center for Biomedical Engineering
Departments of Biomedical & Mechanical Engineering, The School of Engineering of The
City College and The Graduate School of The City University of New York, New York, NY
10031, U. S. A.
E-mail: scowin@earthlink.net

There are two main types of fluid in bone issue, blood and interstitial fluid. The chemical composition of these fluids varies with time and location in bone. Blood arrives through the arterial system containing oxygen and other nutrients, and the blood components depart via the venous system and the lymphatic system containing less oxygen and reduced nutrition. Within the bone, as within other tissues, substances pass from the blood through the arterial walls into the interstitial fluid. The movement of the interstitial fluid carries these substances to the cells encased in the matrix of the bone tissue and, at the same time, carries off the waste materials from the cells. Bone tissue would not live without these fluid movements. There are two purposes of this contribution The first is to provide a description of blood flow and interstitial fluid flow in living bone tissue. The second is to advocate the development of an interactive, dynamically graphical, computational model of blood and interstitial flow in living bone tissue. Such a model will have many significant clinical, research and educational applications. These applications will be discussed.

1   Introduction

Blood and interstitial fluid have many functions in bone. They transport nutrients to, and carry waste from, the bone cells (osteocytes) buried in the bony matrix. They are involved in the transport of minerals to the bone tissue for storage and the retrieval of those minerals when the body needs them. Interstitial flow is thought to have a role in bone's mechanosensory system. Bone deformation causes the interstitial flow over the bone cell membrane; the shear stress of the flowing fluid is sensed by the cell [1,2,3]. A full physiological understanding of this mechanosensory system will provide insight into the following three important clinical problems: (a) how to maintain the long term stability of bone implants, (b) the physiological mechanism underlying osteoporosis, and (c) how to maintain bone in long-duration space flights and long-term bed rest.

Since one purpose of this work is to describe how these fluid systems work, consideration is limited to cortical bone in the mid-diaphysis of a long bone. Although most of what is described is also applicable to bone tissue at other anatomical sites, the discussion is more concise and direct if this limitation is stipulated.

The majority of the motive force for the blood flow is from the heart, but the contraction of muscles attached to bone and the mechanical loading of bone also contribute to this motive force. The majority of the motive force for the interstitial

45

fluid flow is due to the mechanical loading of bone, but the contraction of muscles attached to bone and the heart also supply some of its motive force. The influence of the mechanical loading of a whole bone on the fluid systems' maintenance of the bone tissue is critical. The fluid flow resulting from the mechanical loading is modeled by the theory of poroelasticity. This theory models the interaction of deformation and fluid flow in a fluid-saturated porous medium. The theory was proposed by Biot [4,5] as a theoretical extension of soil consolidation models developed to calculate the settlement of structures placed on fluid-saturated porous soils. The theory has been widely applied to geotechnical problems beyond soil consolidation, most notably problems in rock mechanics. Certain porous rocks, marbles and granites have material properties that are similar to those of bone [6].

The structure of this contribution is first to describe the anatomy of the bone porosity, then to summarize the blood and interstitial fluid movement in bone as well as the factors that drive these flows and cause changes in the flow patterns associated with diseases, surgery and whole body movement. The final sections describe the features and potential applications of the proposed simulation model.

2   Levels of Bone Porosity Containing Blood and Interstitial Fluid

2.1   The Levels of Bone Porosity

There are three levels of bone porosity containing blood or interstitial fluid within cortical bone and within the trabeculae of cancellous bone. A section of a long bone indicating the vascular structure is shown in figure 1 and a more detailed view of the local bone structure is shown in figure 2. The three levels of bone porosity include the vascular porosity associated with the Volkmann canals (Fig. 2) and the Haversian or osteonal canals (Fig. 2), which are of the order of 20 μm in radii; the lacunar-canalicular porosity associated with the fluid space surrounding the osteocytes and their processes (Fig. 3), which are of the order 0.1 μm in radii; and the collagen-hydroxyapatite porosity associated with the spaces between the crystallites of the mineral hydroxyapatite (order 0.01 μm radius). Figure 4 is a scanning electron micrograph (SEM) showing the replicas of lacunae and canaliculi *in situ* in mandibular bone from a 22 year old male. This micrograph illustrates the intense interconnectivity of the lacunar-canalicular porosity. The total volume of the bone fluid vascular porosity is less than that of the lacunar-canalicular porosity [7]. The movement of the bone fluid in the collagen-hydroxyapatite porosity is negligible because most of the bone water in that porosity is bound by interaction with the ionic crystal [8,9]. This portion of the bone water is considered to be part of the collagen-hydroxyapatite structure. The vascular porosity occupied by bone fluid is the space outside the blood vessels and nerves in the Volkmann and Haversian canals (Fig. 2, Fig. 3). This bone fluid freely exchanges with the vascular

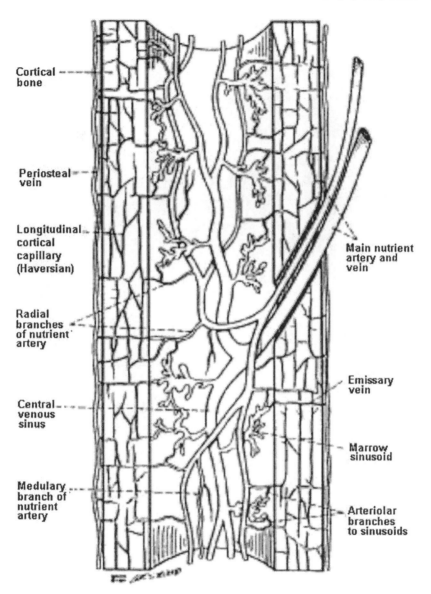

Figure 1. Schematic diagram showing the vascular arrangement in the long bone diaphysis. Modified from Figure 4 of E.A. Williams, R. H. Fitzgerald, and P.J.Kelly, "Microcirculation of Bone," in The physiology and pharmacology of the microcirculation, Vol. 2, Academic Press, (1984).

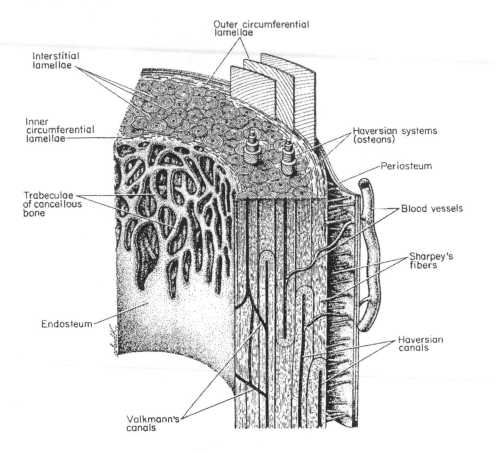

Figure 2. A detailed view of the structure of a typical long bone. From Figure 2.1 of R. B. Martin, D. B. Burr and N. S. Sharkey, Skeletal Tissue Mechanics, Springer, (1998).

fluids because of the thin capillary walls of the endothelium, the absence of a muscle layer, and the sparse basement membrane. There is both outward filtration due to a pressure gradient and inward reabsorption due to osmotic pressure. The function of these flows is to deliver nutrients to, and remove wastes from, the osteocytes housed in the lacunae buried in bone matrix (Fig. 3). An osteocyte left in vitro without nutrient exchange for 4 hours will die [10]. This observation makes sense given the estimate that osteonecrosis in vivo is significant if bone is ischemic for 6 hours or more [11].

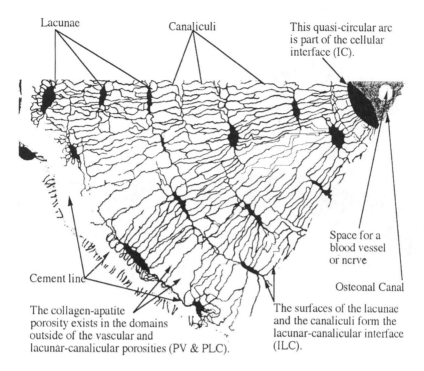

Lacunae

Canaliculi

This quasi-circular arc is part of the cellular interface (IC).

Space for a blood vessel or nerve

Osteonal Canal

Cement line

The collagen-apatite porosity exists in the domains outside of the vascular and lacunar-canalicular porosities (PV & PLC).

The surfaces of the lacunae and the canaliculi form the lacunar-canalicular interface (ILC).

Figure 3. A transverse cross-section of a pie-shaped section of an osteon. The osteonal canal is on the upper right, the cement line to the left. The osteonal canal is part of the vascular porosity, the lacunae and the canaliculi are part of the lacunar-canalicular porosity and the material in the space that is neither vascular porosity nor lacunar-canalicular porosity contains the collagen-apatite porosity. The three interfaces, the cement line, the cellular interface and the lacunar-canalicular interface are each indicated. The radius of an osteon is usually about 100 μm to 150 μm, and the long axis of a lacuna is about 15 μm. Using this information it should be possible to establish the approximate scale of the printed version of this illustration.

2.2   Tracing the Fluid Flow

Over the last thirty years many researchers have used tracers to document bone fluid transport [7,12,19]; see Knothe Tate [19] for a summary of the tracers employed. These tracers show that the normal bone fluid flow is from the marrow cavity to the periosteal lymphatic vessels through the Volkmann and Haversian canals. The flow passes from the Haversian canal into the lacunar-canalicular porosity to the cement line of the osteon, and possibly beyond.

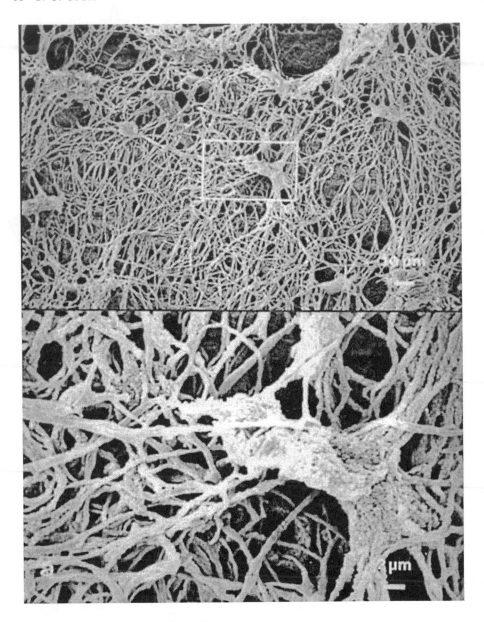

Figure 4. A SEM showing the replicas of lacunae and canaliculi *in situ* in mandibular bone from a young subject aged 22 years. The inset is enlarged below. From P.J.Atkinson and A.S.Hallsworth, (1983) "The changing structure of aging human mandibular bone," *Gerodontology* 2, 57(1983).

2.3   The Fluid Pressures in Long Bones, the BP, ISFP and IMP

Since the blood is encased in blood vessels that are contained within the vascular porosity, the interstitial fluid pressure (ISFP) is less than the blood pressure (BP). The difference between the BP and the ISFP is the transmural pressure. The vascular porosity is a vast low-pressure reservoir for interstitial fluid that can interchange that fluid with the lacunar-canalicular porosity. This is the case because the lineal dimension associated with the bone fluid vascular porosity is two orders of magnitude larger than the lineal dimension associated with the lacunar-canalicular porosity, and the ISFP in the vascular porosity is typically lower than the BP within the blood vessels.

The intramedullary pressure (IMP) in normal bone is the pressure of blood in a local pool of hemorrhage from ruptured intraosseous vessels by drilling into the marrow cavity through the cortex to insert a steel cannula through which the marrow cavity pressure is measured. This was pointed out by Azuma [20] and supported by Shim *et al.* [21]. Therefore, the measurable marrow cavity pressure varies to some extent by the size and type of vessels ruptured as well as by the vasomotor action in the marrow cavity under a condition of anticoagulation. Shim *et al.* [21] noted the differences of the IMP from region to region in a given bone, from bone to bone, and from animal to animal in the same and different species. If the femoral vein was occluded, the IMP was elevated and nutrient venous outflow increased, an indication of venous congestion of bone. If the nutrient or femoral artery was occluded, there was an immediate fall in the IMP and a profound decrease in nutrient venous outflow. The IMP can be increased by mechanical loading of the bone and by venous ligature, and the IMP can be reduced by exercise of the muscles of the calf.

2.4   The ISFP and Mechanosensation

Since the ISFP in the porosity with the largest lineal dimension, the vascular porosity, is always low; the middle porosity - the lacunar-canalicular porosity - appears to be the most important porosity for the consideration of mechanical and mechanosensory effects in bone. The ISFP in the lacunar-canalicular porosity can be, transiently, much higher. A detailed theoretical model of the contents of the lacunar-canalicular porosity has been presented [1,2]. The lacunar-canalicular porosity is the primary porosity scale associated with the relaxation of the excess pore pressure due to mechanical loading. This relaxation of the ISFP was illustrated in Wang *et al.* [22]. In this work the ISFP distributions across a bone are calculated using an idealized bone microstructural model consisting of six abutting square osteons with circular osteonal canals (Fig. 5). This idealized model is shown in the top two panels of figure 5; it has a length of 1200 μm and a width of 200 μm. The ISFP profiles are for different conditions of loading and of permeability of the cement line that forms the outer boundary of the osteon. The completely free flow across the osteonal cement line represents 100% coupling of the osteon with its

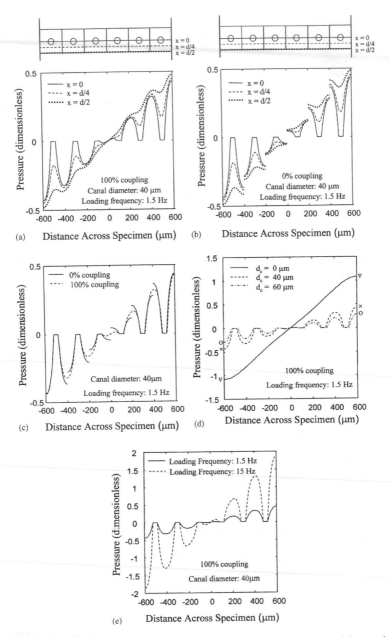

Figure 5. Dimensionless ISFP distributions across a bone using an idealized bone microstructural model. From L.Wang, S.P.Fritton, S.C.Cowin, S.Weinbaum, "Fluid pressure relaxation mechanisms in osteonal bone specimens: modeling of an oscillatory bending experiment", *J. Biomechanics*, 32, 663 (1999).

neighboring osteons, and 0% osteonal coupling is the case in which there is no flow across the cement line. In figure 5(a) and 5(b) the ISFP profiles for a bone model specimen with 40 μm osteonal canals were subjected to an external loading applied at 1.5 Hz for 100% coupling (a) and for 0% osteonal coupling (b). In figure 5(a) and 5(b) the ISFP profiles are plotted along lines whose x distance is expressed as a multiple of the osteonal canal diameter d; the ISFP profiles along the line $x = 0$ are the profiles along a line passing through the canal centers; $x = d/4$ are ISFP profiles along a line halfway between the canal centers and the cement line; $x = d/2$ are ISFP profiles along a line passing through the cement lines. In figure 5(c) the local ISFP gradients for 0% coupling and 100% coupling are compared in the case $x = 0$. In figure 5(d) the effects of the different sized osteonal canals (d =0, 40, or 60 μm) on the pressure profiles and the transcortical ISFP difference are illustrated for 100% osteonal coupling with the loading applied at 1.5 Hz. The transcortical pressure difference is the pressure difference between the points marked "∇", "x" and "O" on the external surfaces. In figure 5(e) the effects of two different loading frequencies, 1.5 and 15 Hz, are illustrated by plotting the transcortical ISFP differences at these frequencies. The effect of frequence on the ISFP gradients is also illustrated by this panel. The lacunar-canalicular porosity is the porosity associated with the osteocytes that are the prime candidates for the mechanosensory cell in bone because of the fluid movement induced by the ISFP gradients [1]. The bone fluid in the smallest porosity, the collagen-hydroxyapatite porosity, is considered to be immovable under normal conditions because it is bound to the collagen-hydroxyapatite structure.

3   Arterial Supply

3.1   Overview of the Arterial System in Bone

All elements of bone, including the marrow, perichondrium, epiphysis, metaphysis, and diaphysis, are richly supplied by vasculature. Mature long bones in all species have three sources of blood supply: (1) the multiple metaphyseal - epiphyseal vessel complex at the ends of the bones, (2) the "nutrient" artery entering the diaphysis (Fig. 1), and (3) the periosteal vessels (Fig. 1, Fig. 2). After entering the diaphysis, the nutrient artery divides into ascending and descending branches, which have further, radially orientated, branches streaming to the bone cortex (Fig. 1). Usually a single nutrient artery enters the diaphysis of a long bone, though many human long bones such as the femur, tibia, and humerus often have two. When the nutrient artery enters bone, the vessel has a thick wall consisting of several cell layers, but within the medulla it rapidly becomes a thin-walled vessel with two cell layers and minimal supporting connective tissue [23]. After reaching the medullary cavity the nutrient artery divides into an ascending and descending branch, which proceed towards the metaphyseal bony ends (Fig. 1). These branches approach the epiphyseal ends of bone, subdividing repeatedly along the way into branches, which pursue a helical course in the juxta-endosteal medullary bone. The terminal branches of the main ascending and descending branches supply the ends of the long bone and anastomose freely with the metaphyseal vessels. The vessels divide and subdivide to feed into a complex network of sinusoids (Fig. 1, Fig. 6). In the immature bone, the open cartilaginous epiphyseal growth plate separates the epiphyseal and metaphyseal vessel complexes.

The blood supply to cortical bone may come from either the medullary canal (younger animals) or the periosteum (older humans, see below). The transcortical blood supply transits in the Volkmann canals and the longitudinal blood supply transits in Haversian systems or osteons. Haversian arteries run longitudinally in osteons (Haversian systems) oriented roughly about 15 degrees to the long axis of a bone. Human cortical bone is largely Haversian at a rather young age compared to other animals. The thin-walled vessels in the cortical canals of Haversian and Volkmann canals are contained in hard unyielding canals in cortical bone and serve to connect the arterioles (the afferent system) with the venules (the efferent system), but unlike true capillaries, they apparently are not able to change diameter in response to physiologic needs [24]. Diffusion from Haversian vessels is insufficient to maintain sufficient nutrition for the osteocytes, convection driven by the ISFP gradients illustrated in figure 5 is necessary for the viability of these cells. Canaliculi serve to connect osteocytic processes [25]. Increased distance from the vascular source (the Haversian artery) probably accounts for the finding that interstitial bone is more susceptible to ischemia than Haversian bone [26].

3.2   Dynamics of the Arterial System

In considering the hemodynamics of any tissue, important elements to be considered are fluid and tissue pressure, fluid viscosity, vessel diameter, and the capillary bed. Blood vessels in bone are richly supplied with nerves and are intimately connected to vasomotor nerve endings; these nerves presumably exert a precise control over blood flow in bone [27]. It is known that in most soft tissues the arteriolar mechanism reduces the BP from 90 mmHg or more in arteries to about 35 mmHg at the arterial end of capillaries. Arterial vessels will close unless the transmural pressure is positive, that is to say, unless the BP in the capillaries does not fall below that in the extravascular space, the ISFP. Note that transmural pressure (BP minus the ISFP) must initially exceed osmotic pressure if filtration is to occur. Absorption of tissue fluid depends upon the transmural pressure being less than the osmotic pressure of the blood at the end of the sinusoid. Osmotic pressure is generally held to be about 20 mmHg. It follows that the pressure in the collecting sinuses of the diaphyseal marrow may be of the order of 55 mmHg. Note that one mmHg is 133.3 Pa or that 3 mmHg is approximately 400 Pa, 80 mmHg is approximately 8 kPa. Bone fluids are interesting in that they exhibit metabolically produced differential diffusion gradients [9,28]. They are sometimes limited in range, but well documented. Thus many ions, such as potassium, calcium, and phosphorus, exist in very different concentrations between blood and bone [29].

3.3 Transcortical Arterial Hemodynamics

Bridgeman and Brookes [30] show that aged bone cortex is supplied predominantly from the periosteum in contrast to the medullary supply in young human and animal material, based on cross-sections through the middiaphysis. They argue that this change is attributed to increasingly severe medullary ischemia with age, brought on by arteriosclerosis of the marrow vessels. They note that an examination of the findings reported by investigators of animal bone blood supply in the past 40 years shows a large measure of agreement. Long standing controversy seems to be based on a failure to recognize that marrow ischemia accompanies natural senescence affecting transcortical hemodynamics and entraining an increasing periosteal supply for bone survival in old age The change over from a medullary to a periosteal blood supply to bone cortex is the consequence of medullary ischemia and reduced marrow arterial pressure, brought about by medullary arteriosclerosis.

3.4 The Arterial System in Small Animals may be Different from that in Humans

The marrow and cortical vascular networks in the rodent are thought to be in series while they are in parallel in the human. From perfusion studies on small mammals (guinea pig, rat, and rabbit) it has been concluded that the blood flow in long bones is such that the major blood supply to the bone marrow is transcortical [31]. This means that the marrow and cortical vascular networks in the rodent are in series. Anatomic and perfusion studies in humans suggest that the circulations of the cortex and marrow are arranged in parallel from a longitudinally running nutrient artery [32]. It was shown that the marrow sinusoids near the endosteal surface of bone typically receive a small arteriolar branch off the major conduit vessel as it enters the bone cortex.

4 Microvascular Network of Marrow

The vascularization of the marrow is illustrated in figure 6. The radial branches of the nutrient artery form a leash of arterioles that penetrate the endosteal surface to supply the bone capillary bed. Small arterioles from these radial branches supply the marrow sinusoids adjacent to the bone. In the adult dog, the marrow consists of adipose tissue (yellow marrow), which provides support for the lateral branches of the nutrient artery as they run toward the endosteal surface of the bone. However, in

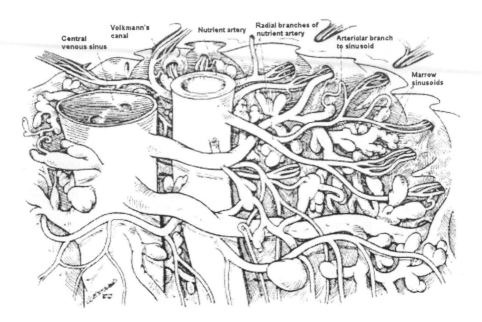

Figure 6. The relationship between the marrow and cortical bone circulations. The radial branches of the nutrient artery form a leash of arterioles that penetrate the endosteal surface to supply the bone capillary bed. Small arterioles from these radial branches supply the marrow sinusoids adjacent to the bone. Modified from Figure 4 of E.A. Williams, R. H. Fitzgerald, and P.J.Kelly, "Microcirculation of Bone," in *The physiology and pharmacology of the microcirculation*, Vol. 2, Academic Press, (1984).

the immature animal, much of the marrow cavity is filled with active hemopoietic tissue (red marrow). The type of capillary varies between red and yellow marrow. Although it is easy to distinguish these types of marrow macroscopically, when seen microscopically, there is no clear-cut separation. The appearance can range from highly cellular to completely fatty. In active red marrow, the small vessels are thin-walled sinusoids, so called because they are many times the size of ordinary capillaries. Despite the thin walls of these vessels, 'I'rueta and Harrison [33] were not able to demonstrate open fenestrations between the endothelial cells. However, it should be noted that Zamboni and Pease [34], using electron microscopy, considered the vessels in red marrow to consist of flattened reticulum cells with many fenestrations and no basement membrane. This would mean that there is minimal hindrance at the sinusoid wall for molecular exchange. In the fatty marrow, the capillaries are closed and continuous like those of other tissues such as muscle [33]. This is supported by in vivo observations in the rabbit [35] that the vessels varied according to the functional state of the marrow. It was estimated that the sinusoid was up to seven times the size of the marrow capillaries, which have a diameter of 8 microns.

5 Microvascular Network of Cortical Bone

Throughout the cortex of long bones there is a capillary network housed in small passages (Fig. 7). In immature bones, these are arranged rather haphazardly, but as bone remodels and matures, a more distinct pattern emerges. In the mature dog and the human, there are two basic systems, the Haversian canals, which run longitudinally, and the Volkmann canals, which run radially (Fig. 2). The two systems are intimately anastomosed to each other. The vessels within the Haversian canals of the human tibia have been examined by microscopy of decalcified sections [36]. The majority of the vessels were observed to be a single layer of endothelial cells. Occasionally, near the endosteal surface of the cortex, small arterioles with a muscular coat were seen, usually accompanied by a larger vein.

A comprehensive examination of the cortical bone of mature and immature dogs by electron microscopy has been reported in Cooper *et al.* [37]. This revealed considerable detail of the capillaries in bone. The Haversian canals ranged in size from 5 to 70 microns and contained either one or two vessels that had the ultrastructure of capillaries. On transverse section, they were lined by one or more endothelial cells, which were surrounded by a continuous basement membrane 400-600 A thick. The junctions of the endothelial cells varied from simple juxtapositioning to a complex interlocking. These investigators found no smooth muscle cells in the walls of the vessels in the Haversian canals. This picture is supported by electron microscopy studies [38] that showed that the cortical capillaries of the growing rat were similar to those found in skeletal muscle, although a basement membrane surrounding the capillaries could not be demonstrated. Thus it appears that the capillaries of bone are a closed tube formed from a single layer of endothelial cells. It has been suggested that transendothelial passage of substances involves two separate pathways; one through the intercellular clefts for hydrophilic substances and another across the endothelial cells themselves for lipophilic substances. If the intercellular capillary clefts are present, they are probably filled with material that makes their permeability low. This is suggested by the work of Cooper *et al.* [37], who observed spaces of 175 A between adjacent endothelial cells that were filled with an amorphous material seen by electron microscopy.

6 Venous Drainage of Bone

The venous complexes draining a long bone parallel those of the arteries. Many workers have commented [33,36] on the extreme thinness of their walls. In the marrow, the venous sinusoids drain into a large, single-cell-walled, central venous sinus, which in turn drains into the "nutrient" veins of the diaphysis. In the adult dog, this thin-walled "nutrient" vein accounts for only 10% of the drainage from the diaphysis [39]. The multiple, penetrating, venous radicles in the metaphysis and epiphysis are also thin-walled and run a more tortuous course than the arteries [33].

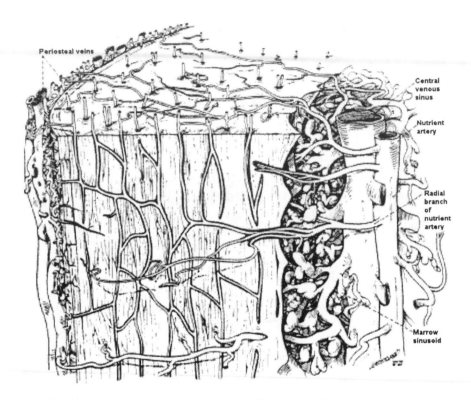

Figure 7. The capillary network within cortical bone. The major arterial supply to the diaphysis is from the nutrient artery. There is an abundant capillary bed throughout the bone tissue that drains outwards to the periosteal veins. Modified from Figure 4 of E.A. Williams, R. H. Fitzgerald, and P.J. Kelly, "Microcirculation of Bone," in The physiology and pharmacology of the microcirculation, Vol. 2, Academic Press, (1984).

The major share of the venous blood leaving long bones has been shown by phlebography to travel by this route [39]. The abundantly anastomosing periosteal network of veins is considered by some workers to drain the diaphyseal bone cortex completely under normal conditions [40]. Many of the veins leaving the long bone pass through muscles, in particular the calf muscle in the case of the lower limb. The alternate contraction and release of the muscles containing the veins is effectively a pump returning the blood towards the heart and away from the bone and decreasing the IMP. The IMP can be reduced by exercise of the muscles of the calf. This arrangement in the case of the calf muscles is illustrated in figure 8. It is clear that the long bone as a whole has multiple venous pathways, the relative importance of which can vary with time and circumstance.

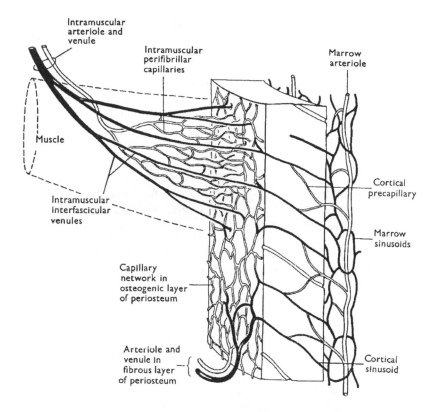

Figure 8. The vascular connection between bone marrow, cortex, periosteum and attached muscle. From Fig. 9.37 of M. Brooks and W.J. Revell, *Blood Supply of Bone*, Springer, London, (1998).

Impaired venous circulation (venous stasis) has been shown to stimulate periosteal bone formation or increase bone mass in the young dog [41], the young goat [42] and in a disuse, hindlimb suspended, rat model [43]. Venous stasis was induced in the experimental animals by applying tourniquets or vein ligation that lasted from 10 days (with additional 30 days recovery [43]) up to 42 days [41] before the bones were examined. There are many other studies demonstrating similar effects [44-47]. An hypothesis for the underlying mechanism of the periosteal bone formation induced by venous stasis has been presented [48].

7   Lymphatics in Bone

The existence of lymphatic vessels in bone remains unclear. On physiologic evidence, some sort of lymph circulation must be present. Large molecules, such as albumin (mol wt 68,000) and horseradish peroxidase (mol wt 40,000), have been shown to leak out of bone capillaries into the interstitial fluid [17,49], and they must have a pathway to return to the general circulation. Kolodny [50] demonstrated that two weeks after India ink was injected into the medullary cavity of long bones, carbon particles were found in the regional lymph nodes.

However, attempts to demonstrate discrete lymphatic vessels within marrow and bone tissue have been consistently unsuccessful. It has been shown with injection studies using thorotrast [12] that this substance leaks from the capillaries of cortical bone into the perivascular fluid and that eventually it can be seen in the periosteal lymphatic vessels. A similar finding has been observed in cortical bone after the use of India ink [51]. The indirect conclusion seems to be that, although there are no demonstrable lymphatic channels in bone tissue, the perivascular fluid as a whole circulates toward the periphery of the bone, carrying with it substances such as large proteins and carbon particles to be taken up by the lymphatics of the periosteum.

Anderson [52] noted that high arterial pressure in the bone marrow probably correlates with an absence of lymphatics in bone marrow and cortex.

8   A Proposal for a Simulation Model

The development of an interactive, dynamically graphical, computational model of blood and interstitial flow in living bone tissue is advocated here. Such a model will have many significant clinical, research and educational applications.

Mechanical loading, BP, IMP, venous ligation, venous pooling and surrounding muscle activity and many chemically driven processes influence the physiologically significant fluid movements in long bones. The dependence of these fluid movements on so many factors makes it very difficult for one to comprehend any particular fluid movement. The following quote from the work [51] of Albert Harris, a biologist, describes the situation: "Systems of interacting forces and stimuli don't have to be very complicated before the unaided human intuition can no longer predict accurately what the net result should be. At this point computer simulations, or other mathematical models, become necessary. Without the aid of mechanicians, and others skilled in simulation and modeling, developmental biology will remain a prisoner of our inadequate and conflicting physical intuitions and metaphors."

9   Features of the Simulation Model Proposed

9.1   Prediction of Blood Flow Redistribution

There are changes in the blood supply to a long bone due to ageing, surgery and injury. The changeover with age from a medullary to a periosteal blood supply to bone cortex is the consequence of medullary ischemia and reduced marrow arterial pressure, brought about by medullary arteriosclerosis. In section 4.3 on transcortical arterial hemodynamics it was noted that Bridgeman and Brookes [30] have shown that aged bone cortex is supplied predominantly from the periosteum in contrast to the medullary supply in young human and animal material. They argue that this change is attributed to increasingly severe medullary ischemia with age, brought on by arteriosclerosis of the marrow vessels. It is clear that if the medullary circulation has not been compromised by age, it will be compromised by an implant that occupies the medullary canal. A desirable property of the proposed model will be to predict these changes.

9.2   Influence of Mechanical Loading

Mechanical loading of the bone can transiently increase the IMP. A desirable property of the proposed model will be to predict these mechanical loading induced changes.

9.3   Influence of Venous Ligature

The IMP can also be increased by application of venous ligature [41-48]. A desirable property of the proposed model will be to predict the IMP increase due to venous ligature.

9.4   Influence of Muscle Contraction

The IMP can be reduced by exercise of the muscles of the calf. A desirable property of the proposed model will be to predict the IMP reduction due to the exercise of the calf muscles.

9.5   Influence of Venous Pooling Capacity

The IMP is influenced by the large venous pooling capacity, 6 to 8 times the arterial volume. The pooling capacity raises the possibility of vessel collapse and ischemia. A desirable property of the proposed model will be to predict influence of the large venous pooling capacity in bone.

9.6   Influence of the Heterogeneity of the Marrow

The contents of the medullary canal is a porous cylinder enclosing globular porous sinusoids, distended fat cells, gel-like extracellular matrices, and a variety of blood components from platelets to neutrophils in various stages of maturity, all of which form a viscoelastic composite. The IMP in the diaphysis is greater than that in the metaphysis or the epiphysis. The IMP is difficult to measure, as was noted in section 2.3 above on the fluid pressures in long bones.

9.7   Influence of Hormonal and Calcium Control Systems

Blood flow is also controlled by non-mechanical mechanisms. Thyroid hormone is known to increase blood flow in dog bones and calcium interchanges: hormonal control of Ca [53].

10   Applications of the Simulation Model Proposed

10.1   Clinical Applications

10.1.1   Ischemia

Ischemia is the term used to describe the insufficient blood supply to a tissue or organ. It is particularly pronounced in bones.

10.1.2   Atherosclerosis in Bone

Atherosclerosis or arteriosclerosis is a degenerative process that affects most blood vessels in most people (in varying degrees) and begins in early teens. Its progression is related in part to genetics, diet, smoking, and high blood pressure. It can be exacerbated by injuries such as trauma or surgery; it causes narrowing and irregular surfaces in blood. It is the most frequent problem in arteries. This degenerative process occurs about 10 years earlier within the bone than elsewhere in the body.

10.1.3   Osteoarthritis

Osteoarthritis is a chronic disorder of a joint with excessive erosion of the cartilage surface associated with excess bone formation at the margins of the joint and gradual loss of function because of pain and stiffness. This may be the outcome of abnormal mechanical forces, such as prior injury, or a systemic disorder, such as rheumatoid arthritis or gout. There is evidence that a deficiency in venous drainage of joint structures may be associated with osteoarthritis [40].

10.1.4   Effect of Implants on Blood Supply

The placement of an implant in bone obviously changes the pattern of blood circulation. When the implant occupies a significant portion of the medullary canal, it can destroy the medullary circulation [30] and require a conversion of the main supply to the bone cortex to come from the periosteum rather than the nutrient artery.

10.1.5 Stress Fractures

Otter *et al.* [54] describe a hypothesis that bone perfusion/reperfusion initiates bone remodeling and the stress fracture syndrome. Recall that stress fractures have been proposed to arise from repetitive activity of training inducing an accumulation of micro-fractures in locations of peak strain. However, Otter *et al.* [54] note that stress fractures most often occur long before accumulation of material damage could occur; they occur in cortical locations of low, not high, strain, and intracortical osteopenia precedes any evidence of microcracks. They propose that this lesion arises from a focal remodeling response to site-specific changes in bone perfusion during redundant axial loading of appendicular bones. They note that IMP's significantly exceeding peak arterial pressure are generated by strenuous exercise and, if the exercise is maintained, the bone tissue can suffer from ischemia caused by reduced blood flow into the medullary canal and hence to the inner two-thirds of the cortex. Site specificity is caused by the lack, in certain regions of the cortex, of compensating matrix-consolidation-driven fluid flow that brings nutrients from the periosteal surface to portions of the cortex. Upon cessation of the exercise, re-flow of fresh blood into the vasculature leads to reperfusion injury, causing an extended no-flow or reduced flow to that portion of the bone most strongly denied perfusion during the exercise. This leads to a cell-stress-initiated remodeling which ultimately weakens the bone, predisposing it to fracture. The testing of this hypothesis could be done with the proposed model.

10.2 Research Applications

The research applications of the proposed model are numerous: for example, the prediction of the redistribution of blood supply, the prediction of the IMP, the ISFP due to changes in mechanical loading, BP, venous ligation, venous pooling and surrounding muscle activity. The model can be used to aid in the discovery of the nature of bone lymphatics and bone innervations. It was pointed out in section 8 above that the existence of lymphatic vessels in bone remains unclear. On the basis of physiologic evidence there must be some sort of lymph circulation present in bone, but how the circulation is accomplished is a mystery. With respect to bone innervation, it is known from dissections that numerous nerve fibers accompany the afferent vessels in bone, and that many nerve fibers in bone marrow are sensible to painful stimuli including distension. If the proposed model were specialized for various animals, for example the rat and the canine as well as the human, then a basis for converting experimental data determined from the rat and the canine to the human could be established.

10.3   Educational Applications

The educational ability of the proposed model is somewhat proportional to the effectiveness of the graphical representations of the computationally predicted evolution of the biological processes and the "user friendliness" of the input control process. Assuming that these are both done well, then one can foresee the model being used to train and educate surgeons and to educate biologists and biomechanical engineers.

Acknowledgments

NIH, NSF and the PSC-CUNY Research Award Program of the City University of New York supported this work.

References

1. Weinbaum S., Cowin S.C. and Yu Zeng, A model for the fluid shear stress excitation of membrane ion channels in osteocytic processes due to bone strain. In *Advances in Bioengineering* ed. by R. Vanderby Jr., Am. Soc. Mech. Engrs. New York (1991) p. 718, & Weinbaum S., Cowin S.C., and Yu Zeng, A model for the excitation of osteocytes by mechanical loading-induced bone fluid shear stresses, *J. Biomechanics* **27** (1994) pp. 339-360.
2. Cowin S., Weinbaum S. and Yu Zeng, A case for bone canaliculi as the anatomical site of strain-generated potentials, *J. Biomechanics* **28** (1995) pp.1281-1296.
3. Cowin S.C. and Moss M.L., Mechanosensory mechanisms in bone, In *Textbook of Tissue Engineering* ed. by R. Lanza,   R. Langer, and W. Chick, (2nd edition),(Academic Press, San Diego 2000) pp. 723-738.
4. Biot M.A., Le problème de la consolidation des matières argileuses sous une charge, *Ann. Soc. Sci. Bruxelles* **B55** (1935) pp.110-113.
5. Biot M.A., General theory of three-dimensional consolidation, *J. Appl. Phys.* **12** (1941) pp.155-164.
6. Cowin S.C., Bone Poroclasticity, *J. Biomechanics* **32** (1999) pp. 218-238.
7. Morris M.A., Lopez-Curato J.A., Hughes S.P.F., An K.N., Bassingthwaighte J.B., and Kelly P.J., Fluid spaces in canine bone and marrow, *Microvascular Res.* **23** (1982) pp. 188-200.
8. Neuman W.F., Toribara T.Y., and Mulrvan B.J., The Surface Chemistry of Bone. VII The Hydration Shell (1953).
9. Neuman W.F., and Neuman M.W., *The Chemical Dynamics of Bone*, (University of Chicago Press, Chicago, 1958).
10. James J., Steijn-Myagkaya GL, Death of osteorytes: Electron microsrcopy after in vitro ischaemia, *J. Bone Jt. Surg.* **68B** (1986) pp. 620-624.

11. Catto M., Pathology of aseptic bone necrosis, In *Aseptic Necrosis of Bone* ed. by J.K.Davidson (American Elsevier 2, New York, 1976).
12. Seliger W.G., Tissue fluid movement in compact bone, *Anat. Rec.* **166(2)** (1970) pp. 247-255.
13. Dillaman R.M., Movement of Ferritin in the 2 Day-Old Chick Femur, *Anat. Rec.* **209** (1984) pp. 445-453.
14. Tanaka T., and Sakano A., Differences in permeability of microperoxidase and horseradish peroxidase into the alveolar bone of developing rats, *J Dent Res.* **64** (1985) pp. 870-876.
15. Li G., Bronk J.T., An K.N., and Kelly P.J., Permeability of cortical bone of canine tibiae, *Microcirculation Res.* **34** (1987) pp. 302-310.
16. Simonet W.T., Bronk J.T., Pinto M.R., Williams E.A., Meadows T.H., and Kelly P.J., Cortical and Cancellous bone: Age-related changes in morphological features, fluid spaces, and calcium homeostasis in dogs, *Mayo Clinic Proc.* **63** (1988) pp. 154-160
17. Doty S.D., and Schofield B.H., Metabolic and structural changes within osteocytes of rat bone, In *Calcium, Parathyroid Hormone and the Calcitonins* ed. by Talmage and Munson, eds., (Excerpta Medica, Amsterdam, 1972) pp. 353-364.
18. Doty S.D., Private communication, (1997).·.
19. Tate M.L., Niederer P., Knothe, Knothe U., *In vivo* tracer transport through the lacunocanalicular system of rat bone in an environment devoid of mechanical loading, *Bone* **22** (1998) pp.107-117.
20. Azuma H., Intraosseous pressure as a measure of hemodynamic changes in bone marrow, *Angiology* **15** (1964) pp. 396-406.
21. Shim S., Hawk H., and Yu W., The Relationship between blood flow and marrow cavity pressure of bone, *Surgery, Gynecology and Obstetrics* **135** (1972) pp. 353-360.
22. Wang L., Fritton S.P., Cowin S.C. and Weinbaum S., Fluid pressure relaxation mechanisms in osteonal bone specimens: modeling of an oscillatory bending experiment, *J. Biomechanics* **32** (1999) pp. 663-672.
23. Yoffey J.M., Hudson G., Osmond D.G., The lymphocyte in guinea-pig bone marrow, *J. Anat.* **99** (1965) pp. 841-860.
24. Rhinelander F.W. and Wilson J.W., Blood supply to developing, mature and healing bone. In *Bone in Clinical Orthopaedics* ed. by G. Summer-Smith (W. B. Saunders, Philadelphia 1982) pp. 81-158.
25. Doty S.B., Morphological evidence of gap junctions between bone cells, *Calcif. Tissue Int.* **33** (1981) pp. 509-512.
26. Kornblum S.S., The microradiographic morphology of bone from ischaemic limbs, Graduate thesis, University of Minnesota (1962).
27. Herskovits M.S., Singh I.J., and Sandhu H.S., Innervation of bone. In *Bone: A Treatise, Vol. 3: Bone Matrix*, ed. by B.K.Hall, (Telford Press, New Jersey 1991) pp. 165-185

28. Martin G.R., Firschein H.E., Mulryan B.J., and Neuman W.F., Concerning the mechanisms of action of parathyroid hormone. II. Metabolic effects, *J. Am. Chem. Soc.* **80** (1958) pp. 6201-6204.
29. Neuman M.W., Bone Blood Equilibrium, *Cal. Tissue Int.* **34** (1982) pp. 117-120.
30. Bridgeman G.and Brookes M., Blood supply to the human femoral diaphysis in youth and senescence, *J. Anat.* **188** (1996) pp. 611-621.
31. De Bruyn P.P.H., Breen P.C., Thomas T.B., The microcirculation of the bone marrow, *Anat. Rec.* **168** (1970) pp. 55-68.
32. Lopez-Curto J.A., Bassingthwaighte J.B., Kelly P.J., Anatomy of the microvasculature of the tibial diaphysis of the adult dog, *J.Bone Joint Sur.* **62** (1980) pp. 1362-1369.
33. 'I'rueta J. and Harrison M.H.M., The normal vascular anatomy of the femoral head in the adult man, *J. Bone Joint Surg.* **35B** (1953) 442-461.
34. Zamboni L. and Pease D.C., The vascular bed of red bone marrow, *J. Ultrastruct Res.* **5** (1961) pp. 65-73.
35. Branemark P.-I., *Angiology* **12** (1961) pp. 293-305.
36. Nelson G.E. Jr., Kelly P.J., Peterson L.F.A., Janes J.M., Blood supply of the human tibia, J.Bone Joint Surg. 42A (1960) pp. 625-636.
37. Cooper R.R., Milgram J.W., Robinson R.A., Morphology of the osteon: an electron microscopy study, *J.Bone Joint Surg.* **48A** (1966) pp. 1239-1271.
38. Hughes S. and Blount M., The structure of capillaries in cortical bone, *Ann. R. Coll. Surg. Engl.* **61** (1979) pp. 312-315.
39. Cuthbertson E.M., Siris E., and Gilfillan R.S., The Femoral Diaphyseal Medullary Venous System As A Venous Collateral Channel In The Dog, *J. Bone Joint Surg. Am.* **47** (1965) pp. 965-974.
40. Brooks M. and Revell W.J., *Bloods Supply of Bone* (Springer, London 1998).
41. Kelly P.J.and Bronk J.T., Venous pressure and bone formation, *Microvasc. Res.* **39** (1990) pp. 364-375.
42. Welch R.D., Johnston,2nd C.E., Waldron M.J., Poteet B., Bone changes associated with intraosseous hypertension in the caprine tibia, *J Bone Joint Surg.* Am. **75** (1993) pp. 53-60.
43. Bergula A.P., Huang W., Frangos J.A., Femoral vein ligation increases bone mass in the hindlimb suspended rat, *Bone* **24** (1999) pp. 171-177.
44. Lilly A.D. and Kelly P.J., Effects of venous ligation on bone remodeling in the canine tibia, *J Bone Joint Surg.* Am **52** (1970) pp. 515-520.
45. Arnoldi C.C., Lemperg R., Linderholm H., Immediate effect of osteotomy on the intramedullary pressure in the femoral head and neck in patients with degenerative osteoarthritis, *Acta Orthop. Scand.* **42** (1971) pp. 454-455.
46. Green N.E. and Griffin P.P., Intra-osseous venous pressure in Legg-Perthes disease, *J Bone Joint Surg. Am.* **64** (1982) pp. 666-671.

47. Liu S.L. and Ho T.C. , The role of venous hypertension in the pathogenesis of Legg-Perthes disease. A clinical and experimental study, *J Bone Joint Surg. Am.* **73** (1991) pp. 194-200.
48. Wang L., Fritton S.P., Weinbaum S., Cowin S.C., On bone adaptation due to venous stasis, *J. Biomechanics* **36** (2003) pp. 1439-1451.
49. Owen M.and Triffitt J.T., Extravascular albumin in bone tissue, *J. Physiol.* **257** (1976) pp. 293-307.
50. Kolodny A., *Arch. Surg (Chicago)* **ll** (1925) pp. 690-707.
51. Harris A.K., Multicellular mechanics in the creation of anatomical structures. In *Biomechanics of Active Movement and Division of Cells*, ed. by N. Akkas, (Springer Verlag 1994) pp. 87-129.
52. Anderson D.W., Studies of the lymphatic pathways of bone and bone marrow, *J. Bone Joint Surg.*, **42A** (1960) pp. 716-717.
53. Williams E.A., Fitzgerald R. H., and Kelly P.J., Microcirculation of Bone. In *The physiology and pharmacology of the microcirculation*, Vol. 2, (Academic Press 1984).
54. Otter M.W., Qin Y.X., Rubin C.T., McLeod K.J., Does bone perfusion/reperfusion initiate bone remodeling and the stress fracture syndrome?, *Medical Hypotheses* **53(5)** (1999) pp.363-368.

AN ANALYSIS OF THE PERFORMANCE OF MESHLESS METHODS IN BIOMECHANICS

M. DOBLARÉ, E. CUETO, B. CALVO,
M. A. MARTÍNEZ, J. M. GARCIA, E. PEÑA
Group of Structural Mechanics and Material Modelling (GEMM).
Aragón Institute of Engineering Research (I3A).
Betancourt Building. María de Luna, 5.
E-50018 Zaragoza, Spain.
E-mail: mdoblare@posta.unizar.es

In this paper we analyse the convenience and possible advantages of using meshless methods and particularly the so-called Natural Element Method (NEM) in numerical simulations in Biomechanics. While Finite Elements have been the universal tool during the last decades to perform such simulations, a recently developed wide family of methods, globally coined as meshless methods, has emerged as an attractive choice for an increasing variety of problems. They present some key advantages such as the absence of a mesh in the traditional sense, particularly important in domains of very complex geometry, or their ability to easily handle finite strains and large displacements in a Lagrangian framework. This is due to their relatively less sensitivity to the point distribution. It is well known that Finite Element Methods fail when the simulated process involves high distortion of the computational mesh. In this case it is necessary to employ a remeshing strategy in order to alleviate these errors. But remeshing is a very time-consuming task and a quality three-dimensional mesher is not available for all-purpose applications. In this work we discuss the use of one of the meshless methods, the Natural Element Method, in Biomechanics. This is a field in which very irregular geometries and large strains often appear. After a brief review of this new meshless approach, such a choice is justified and some examples showing the performance of the method are presented.

1 Introduction

Numerical simulation plays a fundamental role in many branches of science. Computational Biomechanics is one of these branches in which the numerical simulation of very complex processes takes place. Since the appearance of the Finite Element Method (F.E.M.) in the fifties it has been undoubtedly the most extended tool to perform such simulations in that field. However, the high complexity of the geometries involved (bones, soft organs, *etcetera*) and, frequently, the large deformations that usually appear make the aspects related to the mesh definition and control an important fact to be considered. Mesh generation in a general case (three-dimensional models suffering finite strains, ...) is far from being completely automatised and the development of a specific finite element model usually takes a large amount of user time. Quality three-dimensional mesh generators are not currently available for very complex geometries such as living organs. Mesh generation usually relies in a large amount of user time and indeed when the modelled organ suffers large deformations, a frequent remeshing strategy is needed in order to avoid numerical errors that can break out the simulation.

In the last years (especially after the appearance of the pioneer work of Nayroles, Touzot and Villon [1] in 1992) we have assisted to the rapid growth of a new family of numerical methods globally coined as *meshless* or *meshfree*. They are mainly based on Galerkin schemes, although there also exist collocation-based approaches. Their main characteristic is the no need of a mesh in the traditional sense. Instead, the connectivity between nodes is generated in a process transparent to the user, thus alleviating the burden associated to the mesh generation. They are also less sensitive to node distributions than the FEM, allowing stronger distortions without losing the local accuracy.

During this last decade, these methods have evolved rapidly, solving many of their initial problems, such as accuracy, imposition of essential boundary conditions, numerical integration, stability, ... Today they constitute an appealing choice for many applications. In this paper we try to prove that the case of biomechanical simulations is one of these fields.

Firstly, we briefly review the family of meshless methods, considering their common aspects as members of the so-called *Partition of Unity* methods [2,3], with special emphasis on the most extended meshless methods, based on the use of *moving least squares* interpolation. Especial attention is paid to the drawbacks of this interpolation technique. We then introduce the method that is more suitable, in our opinion, to its application to Biomechanics. It is known as Natural Element method (N.E.M.) or, globally, as Natural Neighbour Galerkin method [4,5].

In the rest of the paper we present some examples that justify our choice. These concern the simulation of bone internal remodelling, of the knee menisci behaviour under normal loads and of the tendon behaviour considered as quasi-incompressible hyperelastic material.

1.1   Partition of Unity Methods

It is not easy to give a unifying perspective of meshless methods, provided that many of different nature can be classified among them. One of the most complete and up-to-date reviews on the topic has been performed by Li and Liu [6]. Babuška and co-workers have also provided a global view of these

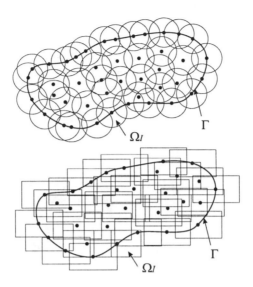

Figure 1. Covering of the domain by circular or rectangular patches.

methods [7]. In this work, the *Partition of Unity* approach is considered as an appropriate unifying framework. Consider the domain of interest Ω discretised into open sets Ω_i (see Figure 1) such that $\overline{\Omega} \subset \cup_{i=1}^{n} \Omega_i$. Also consider a set of basis functions $\varphi_i(\boldsymbol{x})$, whose support is precisely the patch Ω_i.

We say that the functions φ_i form a partition of unity if

$$\sum_{i=1}^{n} \varphi_i(\boldsymbol{x}) = 1, \quad \forall \, \boldsymbol{x} \in \overline{\Omega} \tag{1}$$

In the FEM the partition of unity is composed by the well-known piecewise polynomial shape functions and the patches Ω_i are composed by those elements sharing a given node (support of the approximation functions). Assume now that we are looking for a certain solution $\boldsymbol{u}(\boldsymbol{x})$. Also assume that we know that a certain set of functions $V_i(\boldsymbol{x})$, associated to the patches Ω_i, is a good approximation of the solution on each patch, $\boldsymbol{u}|_{\Omega_i}$. The main conclusion of the work of Babuška [3] is that the global "finite element" space $V = \sum_i \varphi_i V_i$ inherits the approximation properties of the set V_i (and thus approximates well the solution globally), while maintaining the continuity properties of the partition of unity. Note that in the finite element method the functions $V_i = 1$ or, equivalently, there are no *enrichment* of the polynomial space spanned by the base functions.

The basic idea of meshless methods is to use partition of unity functions that do not rely on a mesh to approximate the data. These are often taken from data fitting or interpolation techniques. As enrichment functions one can choose traditional polynomials or simply no enrichment (as in the Element Free Galerkin method [8]), among a wide field of possibilities.

One of the most extended methods to construct the partition of unity is the so-called *moving least squares* approximation, which is used in the Element Free Galerkin method [8], the h-p Clouds method [9], the Diffuse Element method [1] and, more recently, the Meshless Local Petrov-Galerkin method [10].

This is probably the way to construct a partition of unity that has received more attention from the researchers (see the work of Belytschko [11] and references therein). The approximation to the unknown (usually the displacement field) is constructed as follows:

$$u^h \approx \sum_{i=1}^{n} p_i(\boldsymbol{x}) \boldsymbol{a}_i(\boldsymbol{x}) \tag{2}$$

where $\boldsymbol{p}(\boldsymbol{x})$ is a polynomial basis and $\boldsymbol{a}(\boldsymbol{x})$ a set of unknowns to be determined. Usually the polynomial basis is formed by complete linear or quadratic basis functions, although any function can be added to the basis.

The obtained approximation can be made *local* if we particularize the polynomial around a given point, $\overline{\boldsymbol{x}}$:

$$u^h(\boldsymbol{x}, \overline{\boldsymbol{x}}) \approx \sum_{i=1}^{n} p_i(\overline{\boldsymbol{x}}) \boldsymbol{a}_i(\boldsymbol{x}) \tag{3}$$

The $\boldsymbol{a}$ coefficients are then obtained by performing a weighted least squares fit for the local approximation:

$$J = \sum_{I} w(\boldsymbol{x} - \boldsymbol{x}_I)(u^h(\boldsymbol{x}, \boldsymbol{x}_I) - u(\boldsymbol{x}_I))^2$$

$$= \sum_{I} w(\boldsymbol{x} - \boldsymbol{x}_I) \Big[\sum_{i} p_i(\boldsymbol{x}_I) \boldsymbol{a}_i(\boldsymbol{x}) - \boldsymbol{u}_I \Big]^2, \quad I = 1, \ldots n \tag{4}$$

$$\frac{\partial J}{\partial \boldsymbol{a}} = A(\boldsymbol{x}) \boldsymbol{a}(\boldsymbol{x}) - B(\boldsymbol{x}) \boldsymbol{u} = 0 \tag{5}$$

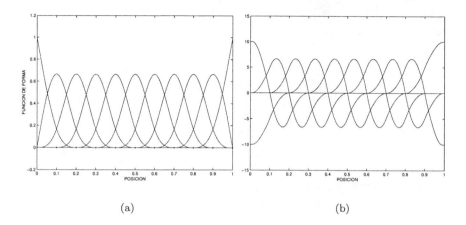

(a)   (b)

Figure 2. Shape functions (a) and their derivatives (b) built with linear basis and support $\alpha = 2.0$.

Finally, in a way more familiar to those involved with Finite Elements, the essential field is approximated as:

$$u^h(x) = \sum_{I=1}^{n} \phi_I^k(x) u_I \qquad (6)$$

being the shape functions given by

$$\phi^k = [\phi_1^k, \phi_2^k, ..., \phi_n^k] = p^T(x) A^{-1}(x) B(x) \qquad (7)$$

Superscript k indicates the order of the approximation —the order of the polynomial exactly reproduced—.

In Figure 2(a) the resultant shape functions in one dimension are depicted, assuming that the support of the shape function covers two nodes. This fact is usually indicated with a parameter $\alpha = supp(\phi_I)/d_I$, where d_I represents the nodal distance. Derivatives of these functions are shown in Figure 2(b).

Extension of these functions to two and three dimensional cases is straightforward either by taking radially supported weighting functions or by tensorial product of one-dimensional shape functions [11].

Methods based on Moving Least Squares have been applied mainly to Solid Mechanics [8], crack propagation in solids [12,13], static and dynamic fracture [11], among others.

Another method that has attracted the attention of many researchers is the so-called *Reproducing Kernel Particle method* (RKPM), both by means of collocation or Galerkin approaches [14].

Essentially, the method begins with the choice of a kernel function $W(\boldsymbol{x}, h)$, where h represents the support of W. With this function acting like a Dirac's delta the strong form of the problem can be *localised* to

$$\boldsymbol{u}^h(\boldsymbol{x}) = \int \boldsymbol{u}(\boldsymbol{y})W(\boldsymbol{x} - \boldsymbol{y}, h)d\Omega_{\boldsymbol{y}} \tag{8}$$

This function W is thus required to have the following properties:

i) $\lim_{h \to 0} \int \boldsymbol{u}(\boldsymbol{y})W(\boldsymbol{x} - \boldsymbol{y}, h)d\Omega_{\boldsymbol{y}} = \boldsymbol{u}(\boldsymbol{x})$

ii) $\int W(\boldsymbol{x} - \boldsymbol{y}, h)d\Omega_{\boldsymbol{y}} = 1$

iii) W has compact support.

iv) W is a function that decreases with distance monotonically.

Equation (8) is then discretised through nodal quadrature, leading to:

$$\boldsymbol{u}^h(\boldsymbol{x}) = \sum_{I:\boldsymbol{x} \in \Omega_I} W(\boldsymbol{x} - \boldsymbol{x}_I)\Delta V_I \boldsymbol{u}_I \tag{9}$$

or, equivalently:

$$\boldsymbol{u}^h(\boldsymbol{x}) = \sum_{I:\boldsymbol{x} \in \Omega_I} \Phi_I(\boldsymbol{x})\boldsymbol{u}_I \tag{10}$$

where Φ_I represent the shape functions of the method.

Since the method thus formulated is known to give unstable results, Liu and co-workers [14] modified the approximation in order to achieve linear precision through the use of a *correction function*:

$$\boldsymbol{u}^h(\boldsymbol{x}) = \int \boldsymbol{u}(\boldsymbol{y})C(\boldsymbol{x}, \boldsymbol{y})W(\boldsymbol{x} - \boldsymbol{y}, h)d\Omega_{\boldsymbol{y}} \tag{11}$$

This method has been also applied to problems in Solid as well as Fluid Mechanics [14] and applied to multiple-scale resolution [15].

Both MLS and RKP methods rely on the use of radial basis functions. They are easy to implement, but lead to a fundamental difficulty on the imposition of essential boundary conditions due to the influence of interior nodes on the boundary. The method presented in the following section uses natural neighbour interpolation to construct the weak form of the problem. This interpolation scheme avoids many of the before mentioned problems and leads to an appealing method for simulations in Biomechanics.

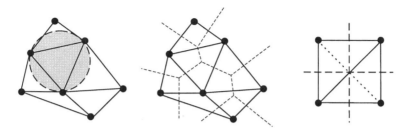

Figure 3. Delaunay triangulation and Voronoi diagram of a cloud of points. On the right, an example of a degenerate distribution of nodes, with the two possible triangulations depicted.

2   Natural Neighbour Galerkin Methods

2.1   The standard NEM

Natural Neighbour Galerkin (or, simply, Natural Element) methods are among the most recent meshless methods. Based upon the early work of Traversoni [16] and Braun and Sambridge [17], Natural Element methods (in what follows, NEM) present unique features among the family of meshless methods. For a deep review on the method and latest developments, see [18]. In essence, NE methods are based on the construction of a partition of unity through the use of the so-called *natural neighbour* interpolation. We present here a brief description of these. For a deeper insight, the interested reader can consult [19,20,21,22,23] for several topics within Solid Mechanics and for some applications in Fluid Mechanics [24,25].

In order to introduce the natural neighbour co-ordinates of a point it is necessary to define two geometrical concepts of first importance, such as the Delaunay triangulation of a cloud of points [26] and its dual structure, the Voronoi diagram [27].

The Delaunay triangulation (tetrahedrisation) of a cloud of points $N = \{n_1, ..., n_N\}$ is the decomposition of the convex hull of the points into k-simplexes (where k represents the dimension of the simplex, that is, $k = 2$ for a triangle and $k = 3$ for a tetrahedron) such that the *empty circumcircle* criterion holds. That is, the circumcircle (circumsphere) of each simplex contains no other point of the cloud N.

The dual structure of the Delaunay triangulation is the decomposition of the space under consideration in the so-called Voronoi diagram of the cloud of points. For a given node n_I, the associated Voronoi cell is composed by all

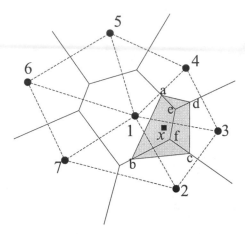

Figure 4. Definition of the Natural Neighbour coordinates of a point x.

of the points which are closer to the node n_I than to any other node:

$$T_I = \{x \in \mathbb{R}^n : d(x, x_I) < d(x, x_J) \ \forall \ J \neq I\} \tag{12}$$

where T_I is the Voronoi cell and $d(\cdot, \cdot)$ represents the Euclidean distance. In the problems considered in this paper, $n = 2, 3$. Following the previous definition, the second order Voronoi cell is defined as the locus of the points that have the node n_I as the closest node and the node n_J as the second closest node:

$$T_{IJ} = \{x \in \mathbb{R}^n : d(x, x_I) < d(x, x_J) < d(x, x_K) \ \forall \ J \neq I \neq K\} \tag{13}$$

Observe that if a new node is added to a given cloud of points, the Voronoi cells will be altered. Sibson [28] defined the natural neighbour coordinates of a point x with respect to one of its neighbours I as the ratio of the cell T_I that is transferred to T_x when adding x to the initial cloud of points to the total area of T_x. In other words, being $\kappa(x)$ and $\kappa_I(x)$ the Lebesgue measures of T_x and T_{xI} respectively, the natural neighbour coordinates of x with respect to the node I is defined as

$$\phi_I^{sib}(x) = \frac{\kappa_I(x)}{\kappa(x)} \tag{14}$$

In Figure 4, this relationship may be written as

$$\phi_1^{sib}(x) = \frac{A_{abfe}}{A_{abcd}} \tag{15}$$

As the principal features of Natural Neighbour Galerkin methods we can say that, unlike many other approximation techniques used in meshless methods, Sibson's interpolation scheme is strictly interpolant, i.e., the approximated surface pass through the data. This can be expressed as

$$\phi_I(\boldsymbol{x}_J) = \delta_{IJ} \qquad (16)$$

with δ_{IJ} the Kronecker delta tensor.

The method also has linear consistency. This can be demonstrated from the before mentioned partition of unity property and its ability to exactly reproduce linear fields (also known as *local coordinate property*, see Sibson [28]):

$$\sum_{I=1}^{n} \phi_I(\boldsymbol{x})\boldsymbol{x}_I = \boldsymbol{x} \qquad (17)$$

In addition, it is easy to demonstrate (see Sukumar [19]) that in two dimensions, Sibson's interpolation leads to constant strain finite element shape functions if the considered point has three neighbours and to bilinear finite element shape functions if it has four neighbours placed on a regular grid. In other cases, Sibson's functions are rational quartic functions.

2.2 Imposition of essential boundary conditions: the α-NEM

Natural neighbour interpolation leads to linear interpolation along the boundary of convex domains. This fact was firstly proved by Sukumar [19], showing that for a point ξ inside a boundary segment 1-2 (see Figure 5) and due to the resulting unbounded area of Voronoi cells associated with nodes 1 and 2 on the convex boundary, the contribution of interior node 3 vanishes along Γ_u and the shape functions associated to the boundary become linear, that is,

$$\phi_1(\xi) = 1 - \xi, \quad \phi_2(\xi) = \xi, \quad \phi_3(\xi) = 0. \qquad (18)$$

This property is, however, not valid for non-convex domains, where the contribution of interior nodes on the boundary is finite; in Sukumar [4] errors of around 2% are reported. Thus, one can conclude that a sampling criterion is required to achieve true interpolation along non-convex boundaries. To accomplish this there are two choices —the use of a CAD-description of the domain [29] or, secondly, one can use the notion of α-shapes to obtain a discrete representation of the domain.

In Cueto *et al.* [22], a modification was introduced in the NEM. This modification is based on the concept of α-shapes. This concept was first

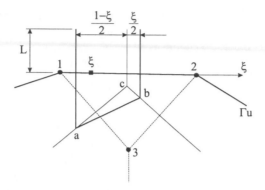

Figure 5. Linear interpolation along a convex boundary Γ_u.

developed by Edelsbrunner [30,31]. It is a generalization of the concept of *convex hull* of a cloud of points and is widely used in the field of scientific visualization and computational geometry to *extract* the shape of a cloud of points.

In essence, an α-shape is a polytope that is not necessarily convex nor connected. It is triangulated by a subset of the Delaunay triangulation of the nodes, and hence the empty circumcircle criterion holds.

If geometrical data is obtained in point-wise manner, in the form of voxels, for instance, these can be treated (usually simplified at lower interest regions) and can be then triangulated in order to obtain an appropriate description of the domain through an appropriate α-shape [31]. In Figure 6 it is seen how the geometry of a human jaw is extracted from nodal data without any prior connectivity. Note also how the α value can be identified with a measure of the "level of detail" up to which the geometry of the domain is represented. Note how it can range from the finer (the cloud itself) to the coarsest level of detail (the convex hull of the cloud of points). This opens the possibility of automatising the extraction of the geometry of biological domains from clouds of points without human intervention.

This definition constitutes the basis for what we have called the α-NEM. If the natural neighbourhood is limited to the case in which two nodes belong to the same triangle (tetrahedron) in a certain α-complex, the linear interpolation property over convex boundaries is also extended to the non-convex case. Hence the Voronoi cells are no longer the basis for the computation of the shape function. Instead, we consider a cell

$$T_I = \{\boldsymbol{x} \in \mathbb{R}^3 : d(\boldsymbol{x}, n_I) < d(\boldsymbol{x}, n_J) \ \forall J \neq I \wedge \sigma_T \in \mathcal{C}_\alpha(N)\}, \qquad (19)$$

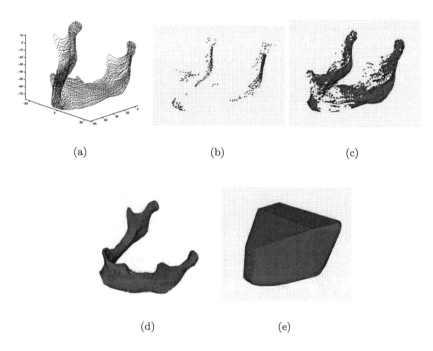

(a) (b) (c)

(d) (e)

Figure 6. Evolution of the family of α-shapes of a cloud of points representing a mandible. Shapes $\mathcal{S}_0$ (a), $\mathcal{S}_{1.0}$ (b), $\mathcal{S}_{1.5}$ (c), $\mathcal{S}_{3.5}$ (d) and $\mathcal{S}_\infty$ (e) are depicted.

where $\mathcal{C}_\alpha(N)$ refers to an appropriate α-complex, and σ_T is the k-simplex that form n_I, n_J and any of the other node in the set N.

Consider a two-dimensional (extension to 3-d is straightforward) regular gridded set N of nodes along a non-convex boundary Γ_u (Fig. 7). Let h be the nodal spacing, and the minimum value α to achieve a proper reproduction of the geometry is $\alpha = \frac{\sqrt{2}}{2}h$. Consider a point x that belongs to Γ_u: for a proper imposition of essential boundary conditions, this point must have an unbounded associated cell. For this to occur, the node A, that in a Delaunay triangulation will make the second order cell associated with x be bounded, must not be a natural neighbour of x. The worst case occurs when x tends to B. The α-ball that would make the points A, B and x pertain to $\mathcal{C}_\alpha(N \bigcup x)$ would have a radius $\alpha' = h$. This is valid whenever the angle formed by the segments $\overline{AB}$ and $\overline{BC}$ is less than $90°$.

In other cases (such as cracks for example), additional information is

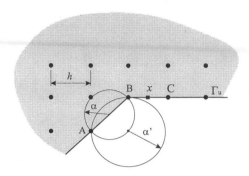

Figure 7. Neighbourhood in the context of α-complexes.

required to fully describe the model. In Sukumar [19], the definition of the domain is based on Planar Straight Line Graph (PSLG), which is a collection of points and segments that must be maintained in the final triangulation. In this case, it would be necessary to use a conforming Delaunay triangulation, and the proper imposition of essential boundary conditions is not guaranteed.

The proposed modification of the shape function computation breaks the duality (only outside the domain) between the Voronoi cells and the triangulation. However, it has been demonstrated that it leads to comparable levels of accuracy on both convex and non-convex domains. For further insight on this topic, consider the situation shown in Figure 8(a), where the Voronoi cell about node 2 is depicted. Assume that the triangle 123 has been subtracted from the Delaunay triangulation in order to meet the α-criterion. Since no neighbourhood is allowed between points placed along segment 1–2 and node 3, the new Voronoi cell becomes unbounded for all points ξ along 1–2; the shape function is shown in Figure 9.

In the authors' opinion, Natural Neighbour Galerkin methods are the best suited for its application to interaction problems due to their ability to exactly interpolate essential boundary conditions [22]. This is of capital importance in Biomechanics since contact problems and interfacial stresses (such as in bone-prosthesis interaction,...) are often an important feature of the problem. Also, the close link between Natural Neighbour Galerkin methods and its geometrical basis (Delaunay triangulations and Voronoi diagrams) make it possible to employ Computational Geometry techniques within a unified framework.

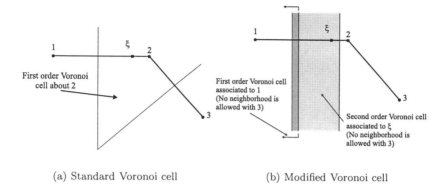

(a) Standard Voronoi cell   (b) Modified Voronoi cell

Figure 8. Effect of the choice of neighbourhood in the Voronoi cell computation.

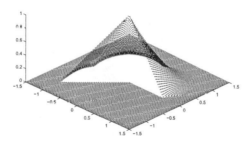

Figure 9. Shape function associated to node B in Figure 7.

2.3   Numerical integration in Natural Element Methods

Numerical integration of the weak form of the problem is a key issue in most meshless methods. This is due to two main factors. On one hand, the employ of background quadrature cells that do not conform to the support of the shape functions and second, the fact that the vast majority (if not all) of the meshless methods are based upon non-polynomial shape functions that also leads to integration errors. Background quadrature cells are necessary in order to apply Gauss or other quadrature schemes for the numerical integration of the weak form of the problem in a Galerkin framework. However, recently, Chen and co-workers [32] have presented a stabilized nodal quadrature scheme with application to moving least squares methods. They adopted a strain smoothing procedure to define the nodal strain operator:

$$\tilde{\varepsilon}_{ij}^h(\boldsymbol{x}_I) = \int_\Omega \varepsilon_{ij}(\boldsymbol{x})\Phi(\boldsymbol{x};\boldsymbol{x}-\boldsymbol{x}_I)d\Omega \tag{20}$$

where ε_{ij} is the strain obtained by compatibility assumptions and Φ is a distribution function. In Chen et al. [32], this function was chosen as

$$\Phi(\boldsymbol{x};\boldsymbol{x}-\boldsymbol{x}_I) = \begin{cases} \frac{1}{A_I} & \text{if } \boldsymbol{x} \in \Omega_I \\ 0 & \text{otherwise} \end{cases} \tag{21}$$

where Ω_I is the Voronoi cell associated with node I and A_I the corresponding area of this cell. With this definition, the strain smoothing leads to

$$\tilde{\varepsilon}_{ij}^h(\boldsymbol{x}_I) = \frac{1}{2A_I} \int_\Omega \left(\frac{\partial u_i^h}{\partial x_j} + \frac{\partial u_j^h}{\partial x_i}\right)d\Omega \tag{22}$$

and on applying the divergence theorem, we obtain

$$\tilde{\varepsilon}_{ij}^h = \frac{1}{2A_I} \int_{\Gamma_I} (u_i^h n_j + u_j^h n_i)d\Gamma \tag{23}$$

where Γ_I is the boundary of the Voronoi cell associated to node I.

This nodal quadrature scheme has rendered excellent results when applied to the RKPM [32], and it is especially well-suited for its application to the NEM, since most of the geometrical entities appearing in its computation (Voronoi cell, circumcenter, etc.) are part of the natural neighbour shape function computations. Hence the additional cost in the implementation of the quadrature scheme is minimal. For a detailed explain on the use of stabilised conforming nodal integration within the NEM framework, the interested reader can consult [33]. This method has shown excellent levels of accuracy.

3   Examples

In this section some examples on the application of the NEM to different biomechanical applications will be shown. They cover different fields, such as hard and soft tissues or prosthetics.

3.1   Application of NEM to the simulation of adaptive bone remodelling

Many computational models of bone adaptation have been developed and implemented in finite element codes [34,35,36,37,38,39,40]. This adaptive process, typical of living tissues, consists of the modification of the bone internal architecture according to the specific mechanical environment. Most of the available models characterise this microstructure by means of macroscopic internal variables. In the case of isotropic models the apparent density or equivalently the porosity are usually the internal variable used [38, 37, 41], while in the case of anisotropic models the internal variables chosen are besides the apparent density, different types of tensors, that characterize the anisotropy in some way [34,36].

These internal variables are treated at the element level and in many cases give rise to discontinuities in the density field, characterized by a "checkerboard" pattern of high and low density elements. Node-based implementations or other type of projection techniques eliminate these kind of numerical problems [42,43,44]. This can also be solved by directly implementing these algorithms with meshless methods with some kind of nodal based integration scheme [18]. Furthermore, in these methods, the support of the shape functions is larger, reducing the problem of localization that often appears in finite element analysis of this class of problems with material softening. We have implemented an anisotropic bone remodelling model, previously developed [34,35], in order to check the ability of this numerical tool to solve the remodelling problem.

This bone remodelling model is based on the principles of Continuum Damage Mechanics (CDM), identifying the local "damage" variable with the bone tissue porosity. Note that this "damage" variable is not related to actual damage processes in the bone matrix as consequence of fatigue or creep, but to the remodelling process.

In fact, in traditional damage models for non-living materials, damage is always positive, as a consequence of the second principle of Thermodynamics, while in living tissues, a metabolic contribution is possible, thus leading to the ability of porosity reduction (or equivalently "damage repair" [34]).

The internal bone architecture is defined with two internal variables: bone apparent density and the so-called "fabric tensor" [46], which characterise respectively the average porosity and the macroscopic directional bone mass distribution. Both variables evolve with time according to the mechanical environment and the principles of CDM, that is, both variables are controlled by stimuli defined, as in standard CDM models, as the thermodynamically associated variables, considering strain as the mechanical independent variable. Finally, after definition of the appropriate bone remodelling criteria, that is, those conditions under which bone resorption and formation mechanisms are activated, the rate of the internal variables is obtained by direct differentiation of the damage criteria (associated law). The interested reader is submitted to the work of Doblaré and García [34] for a complete description of the model.

We now present the results obtained by using this anisotropic bone re-

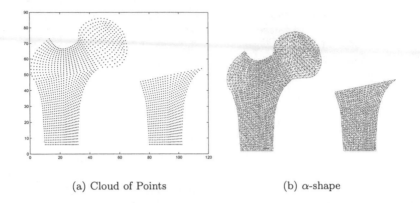

(a) Cloud of Points   (b) α-shape

Figure 10. Cloud of points constituting the input of the problem and associated α-shapes.

Table 1. Load cases, reactions and orientations with respect to the vertical

Load Case	Cycles	Reaction at the Joint	Reaction at the abductor
1	6000	2317 N, 24°	703 N, 28°
2	2000	1158 N, -15°	351 N, -8°
3	2000	1548 N, 56°	468 N, 35°

modelling theory together with a NN approximation of the proximal part of the femur in two dimensions. Two-dimensional models of long bones share the difficulty of not appropriately reproduce the tubular geometry of the diaphisis. Various authors [40,38,36] have proposed the use of a plate that joins these walls, thus leading to a better simulation of cortical bone zones. This plate is assumed in this case to be 5 *mm* thick, while the femur is assumed to be 40 *mm*. They are both joined by means of algebraic constraints (equal displacements of the nodes of both the femur and plate). An α-shape of a femur and its plate, which is generated from the initial cloud of points shown in Figure 10(a), is included in Figure 10(b).

Load is applied in a sequence of three cases, simulating 10000 cycles per time step, as indicated in Table 1, including only the force on the femoral head and the reaction at the abductor due to human walk [47,38,41].

Material has been initially considered homogeneous and isotropic, with a uniform distribution of density, with homogeneous value of 0.5 g/cm^3. The density patterns predicted by the model after 100 and 300 load steps are shown in Figure 11. Numerical integration was carried out by employing three points Hammer quadrature rules over the Delaunay triangles and subsequent strain, stress, bone density and fabric tensor recovered at nodal positions by L_2 recovery [48]. Other possibilities, like stabilised nodal integration techniques, previously employed in Cueto *et al.* [18], are also available.

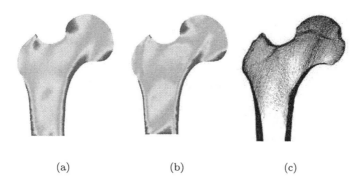

(a)   (b)   (c)

Figure 11. Results of the simulation for 100 days (a) and 300 days (b). Actual radiography of a femur (c).

A great similitude with the morphological structure of the femur (see Figure 11(c)) is noticed. The medullar channel, cortical bone zones at the diaphysis and high density structures corresponding to femoral neck and head can be clearly observed. These results are also highly similar to those obtained by Doblaré and García [34] using Finite Elements (see Figure 12).

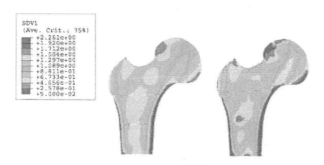

Figure 12. Density patterns for 100 days (left) and 300 days (right) with FEA.

At the same time, we can show in Figure 13 the anisotropic macroscopic distribution in the proximal femur via stress surfaces, that is, the spatial representation of the elastic moduli variation, as a function of the direction has been shown. The anisotropy of the cortical layers is in fact very close to the actual experimental values (see Doblaré and García [34] and references therein). Also, in the neck and head we get a very low degree of anisotropy, while the rest of the epiphysis is almost isotropic. These results are very similar for both numerical methods (FEM and NEM) and the real case.

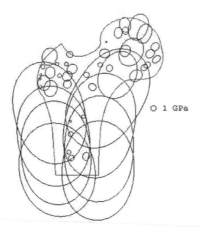

Figure 13. Anisotropic distribution of elasticity modulus in the proximal femur.

Another interesting problem arises, both from the biological and numerical viewpoints, when considering the interaction between two materials. In the case of a hip prosthesis, for instance, the accurate description of stress discontinuities along the bone–prosthesis interface constitutes a numerical challenge for meshless methods. This is because meshless methods, unlike FE, lead to continuously differentiable displacement fields unable to reproduce this discontinuity. However, this is particularly well solved by Natural Elements, as proved in Cueto et al. [23], where it was demonstrated that, by means of the use of appropriate α-shapes of bi-material domains, the discontinuity between materials is accurately reproduced.

Consider, for instance, the same problem presented before, where a total hip replacement has been accomplished (see Figure 14). Once a physiological microstructure has been obtained (see Figure 15), following the approach explained above, the cloud of points is modified in order to consider the in-

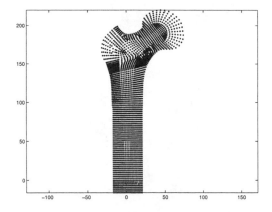

Figure 14. Cloud of points generated for the total hip arthroplasty simulation.

troduction of the prosthesis. Then, the remodelling process is started again.

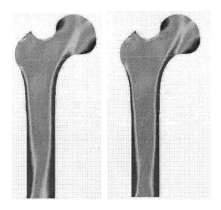

Figure 15. Simulation of the development of a physiological femur structure prior to total hip arthroplasty. 100 days (left) and 200 days (right).

The resulting bone structure after 100 and 250 steps of the hip implantation is shown in Figure 16.

As expected, a bone loss is found near the femoral neck, while a denser structure is developed at the diaphisis. This "stress shielding" phenomenon

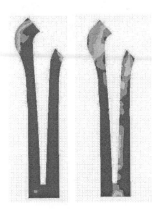

Figure 16. Total hip arthroplasty simulation. Bone density after 100 (left) and 250 (right) days.

is known to be one of the primary causes of failure for this kind of prostheses [49].

3.2   Simulation of hyperelastic tendons under large strains

The main characteristic of biological soft tissues is their high flexibility and anisotropic behaviour determined by the concentration and structural arrangement of their principal constituents: collagen and elastin fibers and the surrounding matrix. Examples of soft tissues are tendons, ligaments, blood vessels, skins or articular cartilages, among many others.

The construction of an accurate constitutive model is difficult because soft tissues are nonlinear, anisotropic, inhomogeneous, viscoelastic, and undergo large deformations. Numerous material models have been developed: elastic, viscoelastic and poroelastic models. Gardiner [50] presents a critical review of the application of different computational techniques to the modelling of ligaments and tendons.

Biological soft tissues are subjected to large deformations and strains with negligible volume changes, that is, only deviatoric or isochoric motions are possible. Such materials can be considered therefore as quasi-incompressible. In addition, they are composed by a matrix material and one or more families of fibers. When one of these family of fibers is predominant, such material has a single preferred direction, that is, it is *transversely isotropic*. For a detailed work on the formulation of a single family of fiber-reinforced hyperelastic materials, see [51,50].

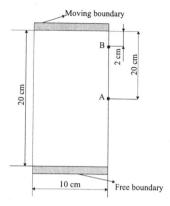

Figure 17. Problem statement for the confined compression test.

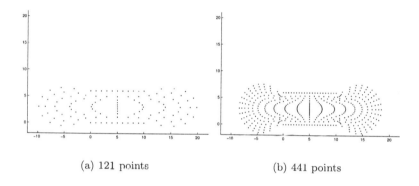

(a) 121 points   (b) 441 points

Figure 18. Deformation for two different number of points.

Some preliminar tests. In order to demonstrate the performance of NE formulations for hyperelastic materials, consider for instance the following two-dimensional plane strain problem [51]. A rectangular rubber component is bonded by two rigid plates on its top and bottom. The problem statement is illustrated in Figure 17. Prescribed displacements are applied to the rigid plates to deform the model up to a vertical nominal strain of 70%. The axial compression is applied on the upper boundary, while the bottom surface remains fixed. The number of points was varied from 121 to 1681 to analyse the model convergence.

Table 2 shows the vertical and horizontal displacements for points A and B (see Figure 17) at maximum deformation state (total vertical displacement of 14 cm) and for the different clouds showing the convergence of the analysis.

Table 2. Displacement for different number of points.

Displacement	211 points	441 points	1681 points	quarter model
$u_x(A)$	9.8114	8.4629	7.6732	7.6960
$u_y(A)$	-6.9803	-7.0053	-6.4910	-7.0
$u_x(B)$	3.1919	2.5546	1.8297	0.7429
$u_y(B)$	-11.5539	-10.8455	-10.097	-10.4452

It is observed in Figure 18 that the lateral sides of the model expand when the number of points increases.

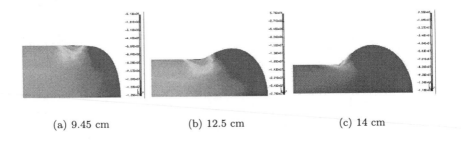

(a) 9.45 cm   (b) 12.5 cm   (c) 14 cm

Figure 19. σ_{33} Cauchy stress contour for natural element simulation.

Commercial code ABAQUS [52] and bilinear quadrilateral, hybrid with constant pressure elements were used in finite element simulations in order to compare with NEM results. With FEM, the maximum displacements until convergence were 12.58, 13.05 and 13.51 cm for the three different meshes before losing convergence, see Figure 20. For NEM, the preestablished displacement of 14.0 cm was achieved for all the point distributions. Consequently, NEM appears to perform better than FEM. This is due to the high distorsion that appears in deformed meshes at large strains, especially near the corners (see Figures 20(d) and 21). Consequently, the stress distributions are not reliable in these regions. In Figure 21 two different intermediate states are presented for the 40×40 mesh of finite element (6.82 and 9.45 cm respectively). The second stress distribution is clearly wrong and only the first case can be considered acceptable. On the contrary, the stress distribution in the NEM is smoother and the results are much better, as seen in Figure 19.

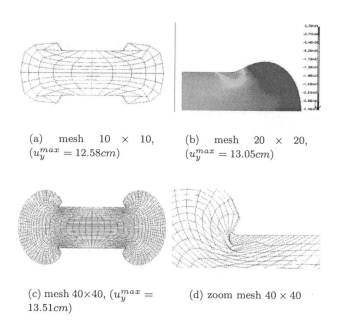

(a) mesh 10 × 10, ($u_y^{max} = 12.58cm$)

(b) mesh 20 × 20, ($u_y^{max} = 13.05cm$)

(c) mesh 40×40, ($u_y^{max} = 13.51cm$)

(d) zoom mesh 40 × 40

Figure 20. Finite element mesh results, until convergence.

A model of the human knee. A detailed model of the human knee is shown in Figure 22. This model includes realistic hard and soft tissue geometries for all the major structures. The surface geometries of the femur, fibula, tibia and patella were reconstructed from a set of Computer Tomography scans, while the different ligaments were obtained from MNR data. Cross-sectional contours were manually digitized from these images and the curves imported into the commercial code I-DEAS [53]. The external surfaces were created by lofting these contours and a regularly distributed cloud of points was generated inside this volume. The femur and tibia were considered as rigid bodies, so only the external surfaces were meshed with rigid "shell" elements. The ligaments were defined by the cloud of points shown in Figure 22(a) and the associated tetrahedrisation in Figure 22(b). An isotropic hyperelastic

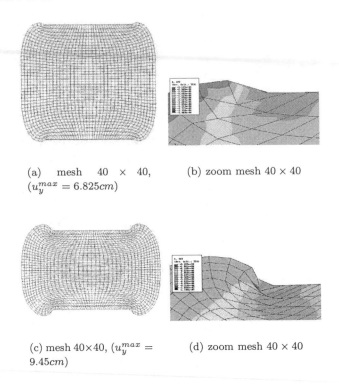

(a)   mesh   40 × 40,   (b) zoom mesh 40 × 40
$(u_y^{max} = 6.825cm)$

(c) mesh 40×40, $(u_y^{max} =$ (d) zoom mesh 40 × 40
$9.45cm)$

Figure 21. Finite element mesh results, realistic stress distribution.

behaviour was used for the ligaments with strain energy function defined as

$$\Psi = \frac{1}{D}(J - 1)^2 + C_1(\bar{I}_1 - 3) \qquad (24)$$

The coefficients were chosen based on material test data [54] and are shown in Table 3. The ligamemnts were attached to the tibia at the distal end and to the femur in its proximal end (Fig. 22(a)). A prescribed displacement and rotation history corresponding to the flexion motion of the knee was applied to the femur until a rotation a 30°. Figure 22(b) shows the maximum principal stress contour for a 30° flexion angle.

The biggest displacement takes place at the MCL, as expected, since this ligament accompanies the femur trough the whole rotation, due to its shape and insertion that covers the inner face of the medial condilus [55].

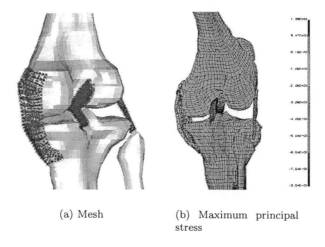

(a) Mesh   (b) Maximum principal stress

Figure 22. Natural element model of the human knee.

Table 3. Material parameters for the Neo-Hookean model for the different ligaments considered (MPa).

Ligament	LCL	MCL	ACL	PCL
Material parrameters	6.06	6.43	5.83	6.06

3.3 Evaluation of the effect of meniscectomies in human knee joint

The meniscus is an important multifuntional component of the knee which has a fundamental role in load transmission, shock absorption, proprioception, improvement of stability and lubrication [56]. Meniscectomy dramatically alters the pattern of static load transmission of the knee joint. Several researchers have noted peak stresses, greater stress concentration and a decreasing in shock-absorbing capability after total or partial meniscectomy [57,58].

In order to investigate the effect of meniscectomy on the human knee joint, a natural element model of the meniscectomized medial menisci were developed for a longitudinal tear.

The geometrical data were obtained from MNR (Magnetic Nuclear Resonance) acquisition. The contours of the distal femur, proximal tibia and menisci were manually digited within each image, (see Figure 23(a)) The material properties were chosen from data available in the literature. Since the stiffness of the bone is much greater than the stiffness of the relevant soft tissues, bones were assumed to be rigid. The meniscus was considered a single-phase material and modelled as isotropic elastic with moduli of $E = 59$ MPa, Poisson ratio of $\nu = 0.45$ [59,60]. A vertical compression force of 1150 N was applied to the upper surface of the femur, corresponding to an average force of the gait cycle for a full extension position [61].

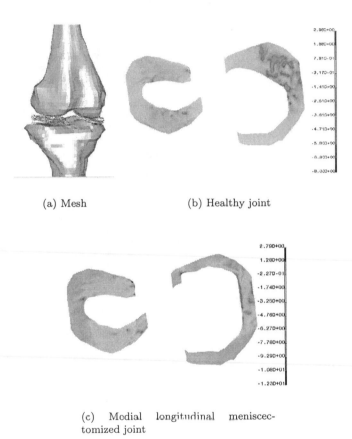

(a) Mesh   (b) Healthy joint

(c) Medial longitudinal meniscectomized joint

Figure 23. Natural element model of the human knee and minimal principal stress contour.

This meniscectomy leads to an increment of the 107% in compression stresses and of the 145% in contact pressures. For the lateral menisci compression stresses reach values of 5.47 MPa. It is important to remark that in this type of meniscectomy, besides the increase of compression, high stresses appear in zones that were initially almost unloaded, changing the fundamental pattern behaviour of the knee and to problems in the long-term, causing damage and remodelling that should be further analyzed.

4 Conclusions

In this paper a study of the possible advantages of using meshless methods in Biomechanical simulations has been performed. It has been shown how meshless methods present some key advantages compared to the Finite Element Method. In particular, no mesh generation is required —it is generated by the method in a process transparent to the user— and there is virtually no limitation to "mesh" distortions, showing that results are much less dependent on the regularity of the nodal distribution than FE methods.

Natural Neighbour Galerkin methods have been chosen by the authors to perform the analysis. This choice has been mainly motivated by two factors. On one hand, Natural Neighbour Galerkin methods are strictly interpolant and thus are very well suited to simulate piece-wise homogeneous domains with high accuracy. On the other, the sound geometrical basis of the NE methods, that has been pointed out, is specially important when dealing with biomechanical structures obtained after volume reconstructions of CT or MRI images, for instance.

Several examples have been presented that show the wide range of problems that NE methods can be applied to. In particular, simulation of bone internal remodelling, hyperelastic modelling of ligaments and the behaviour of the menisci have been studied. Results have been compared to more traditional FE simulations, showing good agreements (even better quality results in some cases). The numerical examples presented in this paper prove the good performance and the high accuracy of the NEM. Compared to the finite element method, NEM can better handle large deformations without any special numerical treatment because of the lower dependence on the original mesh. From these results, the Natural Element Method appears as an efficient alternative to the FEM for this type of problems. This is especially true in three-dimensional simulations where the remeshing processes are a serious drawback, or in applications like in Biomechanics were the original geometry comes from non-CAD based data and the definition of cloud of points is easy to achieve.

Recent works by the authors in other fields such as Fluid Mechanics have leaded us to think that Natural Neighbour Galerkin techniques are able to be successfully applied also in other fields within Biomechanics, such as blood flow, blood–vessel interaction and others.

Acknowledgements

This work was funded by the Spanish Ministry of Science and Technology through grant number CICYT– TIC2000-1935-C04-01

References

1. Nayroles, B., Touzot, G. and Villon, P., Generalizing the finite element method: Diffuse approximation and diffuse elements, *Computational Mechanics* **10** (1992) pp. 307–318.
2. Babuška, I. and Melenk, J. M., The partition of unity finite element method: Basic theory and applications, *Comp. Meth. in Appl. Mech. and Eng.* **4** (1996) pp. 289–314.
3. Babuška, I. and Melenk, J. M., The partition of unity method, *International Journal for Numerical Methods in Engineering*, **40** (1997) pp. 727–758.
4. Sukumar, N., *The Natural Element Method in Solid Mechanics*. PhD thesis, Northwestern University, Evanston, Illinois, (1998).
5. Sukumar, N., Moran, B., Semenov, A. Yu and Belikov, V. V., Natural Neighbor Galerkin Methods, *International Journal for Numerical Methods in Engineering* **50**(1) (2001) pp. 1–27.
6. Li, S. and Liu, W. K., Meshfree and particle methods and their applications, *Applied Mechanics Review* **55** (2002) pp. 1–34.
7. Babuška, I., Banerjee, U. and Osborn, J. E., Meshless and Generalized Finite Element Methods: A Survey of Some Major Results. Technical Report TICAM 02-03, Texas Institute for Computational and Applied Mathematics, University of Texas at Austin, (2002).
8. Belytschko, T., Lu, Y. Y. and Gu, L., Element-Free Galerkin Methods, *International Journal for Numerical Methods in Engineering* **37** (1994) pp. 229–256.
9. Duarte, C.A. M. and Oden, J. T., An H-p Adaptive Method using Clouds, *Computer Methods in Applied Mechanics and Engineering* **139** (1996) pp. 237–262.
10. Atluri, S. N., Kim, H. G. and Cho, J. Y., A critical assesment of the truly Meshless Local Petrov-Galerkin and Local Boundary Integral Equation methods, *Computational Mechanics* **24** (1999) pp. 348–372.
11. Belytschko, T., Krongauz, Y., Organ, D., Fleming, M. and Krysl, P., Meshless methods: An overview and recent developments, *Computer Methods in Applied Mechanics and Engineering* **139** (1998) pp. 3–47.

12. Belytschko, T., Lu, Y. Y. and Gu, L., Crack propagation by element free galerkin methods, *Advanced Computational Methods for Material Modelling* AMD-Vol **180** (1993) pp. 268–280.
13. Fleming, M., Chu, Y. A., Moran, B. and Belytschko, T., Enriched Element-Free Galerkin Methods for Crack Tip Fields, *International Journal for Numerical Methods in Engineering* **40** (1997).
14. Liu, W. K., Jun, S., Li, S., Adee, J. and Belytschko, T., Reproducing kernel particle methods, *International Journal for Numerical Methods in Engineering* **38** (1995) pp. 1655–1679.
15. Liu, W. K. and Chen, Y., Wavelet and multiple scale reproducing kernel methods, *International Journal for Numerical Methods in Fluids* **21** (1995) pp. 901–931.
16. Traversoni, L., Natural Neighbour Finite Elements, In *Intl. Conference on Hydraulic Engineering Software, Hydrosoft Proceedings*, Computational Mechanics publications, (1994) pp. 291–297.
17. Braun, J. and Sambridge, M., A numerical method for solving partial differential equations on highly irregular evolving grids, *Nature* **376** (1995) pp. 655–660.
18. Cueto, E., Sukumar, N., Calvo, B., Cegoñino, J. and Doblaré, M., Overview and recent advances in Natural Neighbour Galerkin methods, *Archives of Computational Methods in Engineering* **10**(4) (2003) pp. 307–387.
19. Sukumar, N., Moran, B. and Belytschko, T., The Natural Element Method in Solid Mechanics, *International Journal for Numerical Methods in Engineering* **43**(5) (1998) pp. 839–887.
20. Sukumar, N. and Moran, B., C^1 Natural Neighbour Interpolant for Partial Differential Equations, *Numerical Methods for Partial Differential Equations* **15**(4) (1999) pp. 417–447.
21. Bueche, D., Sukumar, N. and Moran, B., Dispersive Properties of the Natural Element Method, *Computational Mechanics* **25**(2/3) (2000) pp. 207–219.
22. Cueto, E., Doblaré, M. and Gracia, L., Imposing essential boundary conditions in the Natural Element Method by means of density-scaled α-shapes, *International Journal for Numerical Methods in Engineering* **49-4** (2000) pp. 519–546.
23. Cueto, E., Calvo, B. and Doblaré, M., Modeling three-dimensional piecewise homogeneous domains using the α-shape based Natural Element Method, *International Journal for Numerical Methods in Engineering* **54** (2002) pp. 871–897.

24. Martínez, M. A., Cueto, E., Doblaré, M. and Chinesta, F., Fixed mesh and meshfree techniques in the numerical simulation of injection processes involving short fiber suspensions, *Journal of Non-Newtonian Fluid Mechanics* **115** (2003) pp. 51–78.
25. Martínez, M. A., Cueto, E., Doblaré, M. and Chinesta, F., Natural Element meshless simulation of injection processes involving short fiber suspensions, *Internatonal Journal of Forming Processes* **4**(3-4), (2002).
26. Delaunay, B., Sur la Sphère Vide. A la memoire de Georges Voronoi, *Izvestia Akademii Nauk SSSR, Otdelenie Matematicheskii i Estestvennyka Nauk* **7** (1934) pp. 793–800.
27. Voronoi, G. M., Nouvelles Applications des Paramètres Continus à la Théorie des Formes Quadratiques. Deuxième Memoire: Recherches sur les parallélloèdres Primitifs, *J. Reine Angew. Math.* **134** (1908) pp. 198–287.
28. Sibson, R., A brief description of natural neighbour interpolation, In *Interpreting Multivariate Data. V. Barnett (Editor)*, John Wiley, (1981) pp 21–36.
29. Yvonnet, J., Ryckelynck, D., Lorong, Ph. and Chinesta, F., The meshless constrained natural element method (c-nem) for treating thermal models involving moving interfaces, *International Journal for Numerical Methods in Engineering* to appear, (2003).
30. Edelsbrunner, H., Kirkpatrick, D. G. and Seidel, R., On the shape of a set of points in the plane, *IEEE Transactions on Information Theory* **IT-29**(4) (1983) 551–559.
31. Edelsbrunner, H. and Mücke, E., Three dimensional alpha shapes, *ACM Transactions on Graphics* **13** (1994) 43–72.
32. Chen, J. S., Wu, C. T., Yoon, S. and You, Y., A stabilized conforming nodal integration for Galerkin meshfree methods, *International Journal for Numerical Methods in Engineering* **50** (2001) 435–466.
33. González, D., Cueto, E., Martínez, M. A. and Doblaré, M., Numerical integration in Natural Neighbour Galerkin methods, *International Journal for Numerical Methods in Engineering* to appear, (2003).
34. Doblaré, M. and García, J. M., Anisotropic bone remodeling model based on a continuum damage-repair theory, *Journal of Biomechanics* **35**(1) (2002) pp. 1–17.
35. Bagge, M., A model of bone adaptation as an optimization process, *Journal of Biomechanics* **33**(11) (2000) pp. 1349–1357.
36. Jacobs, C. R., Simo, J. C., Beaupré, G. S. and Carter, D. R., Adaptive bone remodeling incorporating simultaneous density and anisotropy considerations, *Journal of Biomechanics* **30**(6) (1997) pp. 603–613.

37. Weinans, H., Huiskes, R. and Grootenboer, H. J., The Behaviour of Bone-Remodeling Simulation Models, *Journal of Biomechanics* **25** (1992) pp. 1425–1441.
38. Huiskes, R., Weinans, H., Grootenboer, H. J., Dalstra, M., Fudala, B. and Sloof, T. J., Adaptive bone-remodelling theory applied to prosthetic-design analysis, *Journal of Biomechanics* **20**(11/12) (1987) pp. 1135–1150.
39. Beaupré, G. S., Orr, T. E. and Carter, D. R., An approach for time-dependent bone remodeling and remodeling-theoretical development, *Journal of Orthopaedic Research* **8**(5) (1990) pp. 651–661.
40. Beaupré, G. S., Orr, T. E. and Carter, D. R., An approach for time-dependent bone modeling and remodeling-application: A preliminary remodeling simulation, *Journal of Orthopaedic Research* **8**(5) (1990) pp. 662–670.
41. Jacobs, C. R., *Numerical simulation of bone adaption to mechanical loading*, PhD thesis, Stanford University, Stanford, California, (1994).
42. Jacobs, C. R., Levenston, M. E., Beaupré, G. S., Simo, J. C. and Carter, D. R., Numerical instabilities in bone remodeling simulations: The advantages of a node-based finite element approach, *Journal of Biomechanics* **28**(4) (1995) pp. 449–459.
43. Fischer, K. J., Jacobs, C. R., Levenston, M. E. and Carter, D. R., Observations of convergence and uniqueness of node-based bone remodeling simulations, *Annals of Biomedical Engineering* **25** (1997) pp. 261–268.
44. Weinans, H., Huiskes, R. and Grootenboer, H. J., The behaviour of adaptive bone-remodeling simulation models, *Journal of Biomechanics* **25**(12) (1992) pp. 1425–1441.
45. Doblaré, M. and García, J. M., Application of an anisotropic bone-remodelling model based on a damage-repair theory to the analysis of the proximal femur before and after total hip replacement, *Journal of Biomechanics* **34**(9) (2001) pp. 1157–1170.
46. Cowin, S. C., The relationship between the elasticity and tensor and the fabric tensor, *Mech. Mater.* **4** (1985) pp. 137–147.
47. Carter, D. R., Fyhrie, D. P. and Whalen, R. T., Trabecular bone density and loading history: Regulation of tissue biology by mechanical energy, *Journal of Biomechanics* **20** (1987) pp. 785–795.
48. Zienkiewicz, O. C. and Zhu, J. Z., The Superconvergent Patch Recovery and a posteriori error estimates. Part I: the recovery technique, *International Journal forn Numerical Methods in Engineering* **33** (1992) pp. 1331–1364.

49. Huiskes, R., Weinans, H. and Rietbergen, B. van, The relationship between stress shielding and bone resorption around total hip stems and the effects of flexible materials, *Clin Orthop.* **274** (1992) pp. 124–34.
50. Weiss J. A. and Gardiner, J. C., Computational modeling of ligament mechanics, *Critical Reviews in Biomedical Engineering* **29**(3) (2001) pp. 1–70.
51. Martínez, M. A., Calvo, B. and Doblaré, M., Natural Element formulation of hyperelastic materials, *Internatonal Journal for Numerical methods in engineering* submitted), (2002).
52. Hibbit and Karlsson and Sorensen, Inc. *Abaqus user's Manual, v. 6.2.* HKS inc. Pawtucket, RI, USA., (2001).
53. Structural Dynamics Research Corporation. *I-Deas Tutorials.* EDS, (2001).
54. Weiss, J. A., Maker, B. N. and Govindjee, S., Finite element implementation of incompressible, transversely isotropic hyperelasticity, *Computer methods in applied mechanics and engineering* **135** (1996) pp. 107–128.
55. Woo, S. L.-Y., Debski, R. E., Withrow, J. D. and Janaushek, M. A., Biomechanics of the knee ligaments, *Am J Spors Med* **27** (1999) pp. 533–543.
56. Walker,P. S. and Erkman, M. J., The role of the menisci in force transmission across the knee, *Clinical Orthopeadics and Related Research* **109** (1975) pp. 184–192.
57. Fithian, D. C., Kelly, M. A. and Mow, Van C., Material properties and structure-function relationship in the menisci, *Clinical Orthopeadics and Related Research* **252** (1990) pp. 19–31.
58. Fairbank, P. G., Knee joints changes after menisectomy, *J Bone and Joint Surg* **52** (1948) pp. 564.
59. Proctor, C. S., Schmidt, M. B., Kelly, M. A. and Mow, V. C., Material properties of the normal medial bovine meniscus, *Journal of Orthopeadic Research* **7** (1989) pp. 771–782.
60. LeRoux, M. A. and Setton, L. A., Experimental biphasic fem determinations of the material properties and hydraulic permeability of the meniscus in tension, *J Biomech Eng* **124** (2002) pp. 315–321.
61. Sathasivam, S. and Walker, P. S., A computer model with surface friction for the prediction of total knee kinematics, *J. Biomechanics* **30** (1997) pp. 177–184.

ASSESSMENT OF PLAQUE STABILITY BASED ON HIGH-RESOLUTION MAGNETIC RESONANCE IMAGING OF HUMAN ATHEROSCLEROTIC LESIONS AND COMPUTATIONAL MECHANICAL ANALYSIS

CHRISTIAN A.J. SCHULZE-BAUER, MICHAEL STADLER

Institute for Structural Analysis—Computational Biomechanics, Graz University of Technology
Schiesstattgasse 14-B, 8010 Graz, Austria

RUDOLF STOLLBERGER

Abteilung für klinische und Experimentelle Magnetresonanzforschung, Universitätsklinik für Radiologie
Medical University Graz, Auenbruggerplatz 9, 8036-Graz, Austria

PETER REGITNIG

Institute for Pathology, Medical University Graz
Auenbruggerplatz 25, 8036-Graz, Austria

GERHARD A. HOLZAPFEL

Institute for Structural Analysis—Computational Biomechanics, Graz University of Technology
Schiesstattgasse 14-B, 8010 Graz, Austria
E-mail: gh@biomech.tu-graz.ac.at

Background – To date there is no efficient diagnostic strategy available for the assessment of plaque stability. The present work proposes a computational methodology to identify high-risk plaques. The proposed model considers eight major tissue components and allows three-dimensional mechanical analyses of plaque stability subject to various factors such as blood pressure and mechanical properties of tissue components.

Methods and Results – Morphology of an eccentric low-grade lesion was determined by means of multi-phasic and high-resolution magnetic resonance imaging on a 1.5 T whole body system and corresponding histological analyses. Nonlinear anisotropic material models were fitted to experimental data obtained from tensile tests of isolated tissue components. Stress distributions in the lesion were computed using nonlinear finite element analyses. Based on the particular low-grade lesion, the effects of elevated blood pressure were studied, and, additionally, the potential effects of changes in the lipid pool stiffness and the media stiffness, that can be controlled by drug treatment, investigated. Intimal stress distributions are strongly influenced by non-intimal components, and hence suitable modeling of non-intimal and intimal components is equally important. A meaningful stability assessment of individual lesions requires three-dimensional approaches.

Conclusions – Realistic three-dimensional "morpho-mechanical" modeling of atherosclerotic lesions provide information and insights important for the assessment and understanding of plaque stability and related drug effects. This information can not be obtained by entirely morphological approaches.

1   Introduction

There is general agreement that the majority of acute cardiovascular syndromes is caused by vulnerable plaques. Hence, identification of vulnerable plaques is considered a primary objective of vascular diagnostics. Earlier studies showed that plaque cap thickness and lipid core size [1,2,3] are essential determinants of plaque stability. Considering the rapid developments in arterial wall imaging, based mainly on magnetic resonance (MR) imaging (for a review see, for example, [4]) and intravascular ultrasound [5], detection of these micromorphological characteristics appears to be a promising diagnostic strategy. Yet, restriction to morphological criteria is not an approach likely to provide a highly predictive basis for the diagnosis of vulnerable plaques. The reason for this is that, in general, it is impossible to assess the stability of a structure solely by considering its geometry. Stability analysis requires additional information of mechanical loads applied, mechanical properties of involved materials, and the associated mechanical stresses occurring in the structural components.

Previous mechanical studies [1,2,3] have demonstrated the coincidence of plaque rupture sites and regions of circumferential stress concentrations. These landmark studies supported the intuitive hypothesis that the acute process of plaque fracture is caused by high mechanical stresses, which exceed the ultimate tensile strength of the fibrous cap. Consequently, computational mechanics may provide significant contributions for the understanding, identification and treatment of vulnerable plaques. Previous studies, however, utilized simplified modeling approaches based on two-dimensional model plaques. The layered structure of diseased arteries was ignored more or less, and the mechanical behaviors of the considered materials were assumed to be linearly isotropic, which is inconsistent with histology and vascular mechanics [6]. Moreover, the numerical methods employed so far, i.e. mainly linear finite element methods for plane stress states, may also be questionable, in particular, when the blood pressure varies within ranges for which linear theories are not appropriate anymore. Spontaneous hypertension may change the blood pressure from 70 mmHg systole to over 200 mmHg diastole causing a nonlinear stress-strain response in the wall with a typical (exponential) stiffening effect at higher pressures. For supra-physiological loading conditions the simplified models do not yield reliable quantitative results, as documented in [7]. They are restricted to general and rather intuitive insights.

According to [8] the objective of the present work is to design realistic three-dimensional (3D) "morpho-mechanical" models of diseased arteries, which are based on (i) high-resolution MR imaging of human atherosclerotic lesions and (ii) mechanical testing of their isolated tissue components, and to present a computational methodology to identify high-risk plaques. Nonlinear finite element methods are used for the analysis of plaque mechanics. The potential of the proposed approach for the assessment of plaque stability is demonstrated by means of "computational case studies" for one human diseased artery. The present approach allows detailed studies of the mechanical effects of blood pressure elevations and variations of the stiffness of the individual plaque components on the mechanical behavior of the atherosclerotic lesion.

2 Method

2.1 Specimen

An iliac artery was excised from a human cadaver (68 years, male) during autopsy within 24 hours from death. After removal of loose connective tissue the vessel was placed in 0.9 % NaCl solution at room temperature. The ratio of *in situ* length to *ex situ* length, which is known as the axial *in situ* stretch, was measured to be 1.04. Use of autopsy material from human subjects was approved by the Ethics Committee, Medical University Graz, Austria.

2.2 High-Resolution Magnetic Resonance Imaging and Histology

Immediately after excision a diseased segment of the vessel was scanned with a 1.5 T whole body magnetic resonance system (Philips ACS-NT, maximum gradient strength = 23 mT) in 0.9 % NaCl solution at 37° C ± 1° C. For signal reception a circularly shaped surface coil (diameter = 5 cm) was positioned under the container. 3D turbo spin echo sequences were applied to achieve high spatial resolution and a sufficient signal to noise ratio in acceptable scan time (< 20 min). Scan parameters were: Field-of-View = 60 mm, scanning matrix = 256×256, slice thickness = 0.6 mm (zerofilling in z-direction), number of slices 32, echo spacing = 13.5 ms, half Fourier technique (60-80 % k-space). A turbo factor (TF) of 4 was used for T1 (TR/TE = 600/13) and "pseudo proton density" (PD) weighted sequences (TR/TE = 2000/20), and a TF of 10 for T2 weighted sequences (TR/TE = 2000/75).

After image acquisition the segment was cut in two half segments. One half segment was used for component-specific mechanical testing, the other half segment was analyzed histologically for matching with corresponding MR slices. For this purpose the "histological segment" was fixed in 4 % buffered formaldehyde solution, decalcified with ethylene diamine tetraacetic acid (EDTA), embedded in paraffin, and serially sectioned at 0.6 mm intervals. Three micron thick sections were stained with Hematoxylin and Eosin or Elastica van Gieson.

2.3 Segmentation and Geometrical Modeling

Raw MR images were corrected for intensity non-uniformity utilizing the 'N3' method (Nonparametric Non-uniform intensity Normalization) [9] and noise-filtered using a "customized" anisotropic filter (edge/flat/gray filter [10]). Segmentation is performed according to a scheme [7], which distinguishes eight tissue components: adventitia, non-diseased media, diseased (fibrotic) media, non-diseased intima (intimal hyperplasia), fibrous cap, fibrotic intima at the abluminal border (fibrous "bottom"), lipid pool and calcification. Active contour models ("snakes"; for the underlying concepts and current applications in medical image processing see [11]) were used for automatic detection of the outer boundary of the adventitia, the adventitia-media boundary, the media-intima boundary and the intima-lumen boundary. Since calcifications exhibited characteristic dark gray-levels they could be segmented via thresholding. For segmentation of the lipid pool, the fibrous cap, and the non-diseased intima, histological sections of the "histological half" were segmented manually by an histopathologist and matched with multi-phasic MR data. The resulting correlations were used for the segmentation of the intima of the entire vessel segment. Furthermore, histological analysis showed a higher amount of collagenous tissue in the portion of the media adjacent to the plaque. This portion, which differed also in mechanical behavior from the remaining (non-diseased) portion of the media, was labeled as "diseased media".

Based on the segmented MR images 3D geometrical models of the vessel were reconstructed. The component boundaries were described by means of "non-uniform rational B-splines" (NURBS) [12], which provide analytical representations of the structural composition. NURBS are beneficial with respect to mesh refinements for subsequent finite element analyses.

2.4   Mechanical Testing and Physical Modeling

The second half segment intended for mechanical testing was dissected anatomically into its major components according to the segmentation scheme. From all soft tissue components two strips were cut out, one with axial and one with circumferential orientation. Strips underwent quasi-static (crosshead speed: 1 mm/min) and uni-axial extension tests in a tissue bath, whereas axial load and in-plane deformations were recorded continuously. Specimens were preconditioned with five successive loading-unloading cycles. Since calcifications and the lipid pool are inappropriate for the used type of mechanical tests they were excluded from testing. More details on the mechanical experiments and original experimental data are provided in [7,13].

Mathematical representation of the mechanical behaviors of the soft tissue components was based on a large-strain prototype model for fiber-reinforced composites operating in the large strain domain [14]. This physical model captures the characteristic nonlinear, anisotropic and incompressible behavior of soft vascular tissues. Material parameters, documented in [7], were determined by fitting the model to the experimental data using the Levenberg-Marquardt algorithm. Calcifications were modeled numerically as nearly rigid bodies. The lipid pool was represented by an incompressible neo-Hookean model [15] with a shear modulus of 0.15 kPa, which was calculated from data of the literature [1]. Finally, each component of the geometrical vessel model was associated with an individual physical model [15,14,7].

2.5   Numerical Modeling

The physical model was implemented into the finite element program ABAQUS V5.8 by means of the UEL programming interface. Since the tissue components were represented as (nearly) incompressible materials, a three-field variational principle was utilized [15]. The vessel segment investigated was discretized by 3828 eight-node isoparametric brick elements. More details on the nonlinear finite element methods used for the computational analysis may be found in [7].

2.6   Computational Case Studies

Stress distributions in diseased arteries depend strongly on blood pressures, boundary conditions and material properties of their major tissue components. The present study focuses on factors that can be controlled by drug treatment. In particular, the following cases were investigated:

(i)   Variation of the mean arterial pressure (MAP): 95 mmHg, 135 mmHg, 175 mmHg
(ii) Variation of the lipid pool stiffness at an elevated MAP of 175 mmHg: 100 % (reference stiffness), 200 %, and 50 %
(iii) Variation of the media stiffness at an elevated MAP of 175 mmHg: 100 % (reference stiffness), 125 %, and 75 %

For all computational cases the model vessel was axially stretched up to 4 % in order to simulate the axial *in situ* prestretch. Furthermore, the model adventitia was additionally prestretched in axial direction up to 20 %. This adventitial prestretch was observed during surgical dissection of the vessel components [16].

While drug control is evident for blood pressure (item (i)), this is not so clear for the latter two factors. At least in experimental models it was demonstrated that compositional changes in plaques, which were observed during plaque regression in animal experiments, are associated with more than four-fold variations of the elastic shear modulus of model lipid pools [17]. Media stiffness variations were assumed to be ± 25 %. *In vitro* tests of dog iliac arteries showed that the circumferential wall stress is two times higher for norepinephrine-induced maximum vessel contraction than for zero vessel tone (passive state) provided that inflation pressure and vessel configuration remain unchanged [18]. Therefore it is unlikely that the assumed stiffness variation of ± 25 % is too large.

3   Results

Computational analyses provide information on the local multi-axial stress state at each position of the 3D multi-component lesion (Fig. 1). Since autopsy studies have demonstrated that plaque disruptions are oriented mainly along the vessel axis [3], and for the sake of clearness, results presented in this study were restricted to circumferential stresses. Note, however, that, in general, the tensile strength of materials depends on the entire multi-axial stress state.

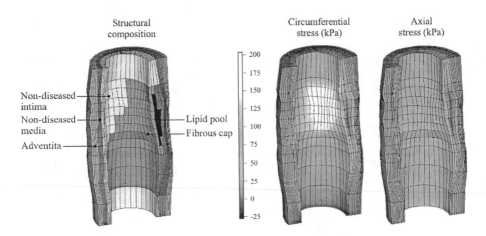

Figure 1. Finite element discretization of the vessel structure (left plot) and plots of Cauchy stresses in circumferential and axial directions for a mean arterial pressure of 95 mmHg.

Stress distributions of the fibrous cap for the considered computational cases are plotted in figure 2, and associated descriptive statistical values such as mean, standard deviation (SD), minimum (Min), and maximum (Max) are listed in Table 1. There were clear changes of the circumferential stress distribution for MAP variations (Fig. 2a), and for the variation of the media stiffness (Fig. 2c). In contrast, stiffness variation of the lipid pool did not yield noticeable changes, as seen from figure 2b.

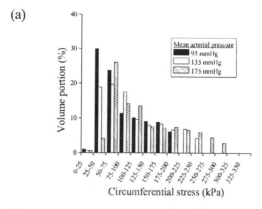

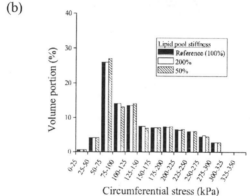

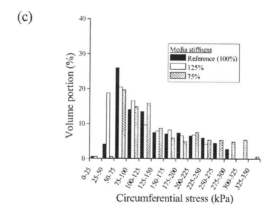

Figure 2. Distributions of the circumferential Cauchy stresses in the fibrous cap for variations of the mean arterial pressure (a), the lipid pool stiffness at 175 mmHg (b), and the media stiffness at 175 mmHg (c).

Table 1.   Descriptive statistics of circumferential stresses in the fibrouscap
for seven computational cases.

Computational case	Circumferential stress in fibrous cap (kPa)			
	Mean	SD	Min	Max
MAP: 95 mmHg	86	48	11	200
MAP: 135 mmHg	107	59	17	246
MAP: 175 mmHg	129	70	23	290
MAP: 175 mmHg Lipid pool stiffness: 200 %	129	70	23	290
MAP: 175 mmHg Lipid pool stiffness: 50 %	129	70	23	290
MAP: 175 mmHg Media stiffness: 125 %	107	59	17	246
MAP: 175 mmHg Media stiffness: 75 %	150	80	29	334

The distributions of the intimal stresses were investigated in more detail and visualized using a Mercator projection (Fig. 3a). This means that the intima was projected radially onto a cylindrical surface with respect to the vessel axis, which was then projected onto a plane. Accordingly, this type of plot shows the "unrolled" intima. The maximum circumferential stresses throughout the thickness of the non-diseased intima and the fibrous cap were plotted. Additionally, for the sake of clearness, the boundaries of the fibrous cap (dotted line), the lipid pool (dashed line), and the calcification (solid lines) were indicated.

In order to illustrate the effect of MAP variations on the stress concentration in the fibrous cap a safety margin was chosen to be about 230 kPa, which is about one third of the circumferential fracture stress measured for the fibrous cap[1]. In figure 3b the 230 kPa-stress contours were plotted for a MAP of 135 mmHg (thick dashed line) and 175 mmHg (thick solid line). For a MAP of 95 mmHg, stresses in the fibrous cap were smaller than 230 kPa, and thus there is no associated stress contour. Within the contours stresses exceed 230 kPa. In an analogous manner the results for the stiffness variation of the media were plotted in figure 3c. In contrast, results for the stiffness variation of the lipid pool were not plotted because the associated stress contours are (nearly) identical.

---

[1] Compared to the fracture stress the safety margin chosen appears to be relatively small. Yet, a small value had to be chosen since the plaque investigated was stable, and thus plaque stresses were relatively small.

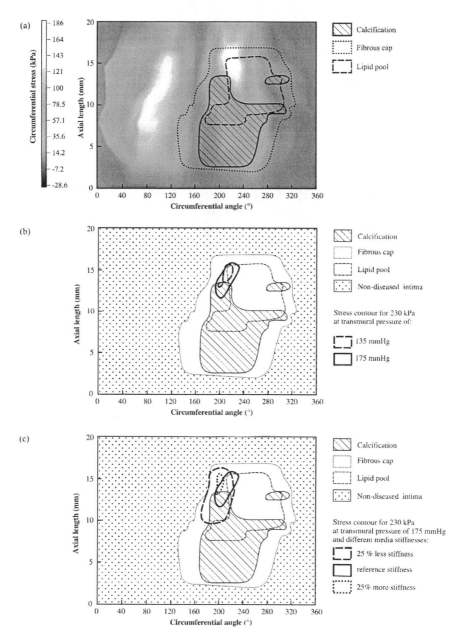

Figure 3. Cauchy stresses in circumferential direction for the "unrolled" intima (Mercator projection) for the mean arterial pressure (a). "Safety margins" (230 kPa stress contours) for different mean arterial pressures (b) and variation of the media stiffness (c). Within the stress contours peak stresses occur, which exceed 230 kPa.

4   Discussion

The proposed computational model seems to be the most advanced approach for arterial wall mechanics to date. In contrast to previous mechanical studies of diseased arteries, the current approach considers the 3D geometry of a real-life human plaque, considers the nonlinear mechanical behavior of eight distinct tissue components from the same individual lesion, and provides fully 3D mechanical analysis (Fig. 1). The present computational model allows the study of the component-specific micro-mechanics of an individual diseased artery subject to various factors, such as blood pressure, geometry and material properties of tissue components etc. It is a valuable computational tool for the analysis of plaque stability and allows investigation of potential drug effects on the changes in the mechanical environment and the associated plaque stability (Figs. 2,3). The computational results demonstrate that plaque stability assessment is a complex mechanical issue, which can hardly be reduced to investigations of simple two-dimensional geometrical features of plaque cross-sections.

4.1   Three-Dimensional Stability Assessment

The complex stress distribution in the intima for a reference MAP of 95 mmHg (Figs. 1 and 3a) is a result of the specific three-dimensional architecture and the nonlinear and anisotropic mechanical properties of the lesion structure. With this approach it is possible to study regions of stress concentrations, which indicate rupture-prone regions in the entire fibrous cap. Computational analyses of this particular atherosclerotic lesion indicate that remarkably, high stresses occur also in the non-diseased intima opposite to the plaque. The associated pathophysiological and clinical meaning have not yet been evaluated. Clearly, the 3D intimal stress pattern can not be anticipated by means of simple geometrical considerations and mechanical intuition. Obviously, also computational cross-sectional analysis fails to capture the complexity of the structure. This may be one of the essential reasons for the problems of earlier studies in determination of the actual rupture sites [3].

4.2   MAP Variation

In addition to the investigation of a physiological reference situation, meaningful stability assessment should consider the influence of important clinical parameters such as blood pressure. Acute cardiovascular syndromes are triggered frequently by elevated blood pressure due to physical activity or emotional stress. The computational model allows detailed studies of the pressure-dependent stress shift towards higher values (Fig. 2a). Information on the variation of stress distributions provides a basis for the assessment of the responsibility for the increase of pressure-induced stresses, and at which pressure level alarming stresses occur. This type of information may help to characterize more precisely the "vulnerability" of a lesion.

Additionally, appropriate visualization allows a direct identification of regions at risk. For example, the encircled regions of figure 3b indicate zones, where local stresses exceed a stress "safety margin" of 230 kPa for transmural pressures of 135 mmHg (thick dashed line) and 175 mmHg (thick solid line). Since the safety margin was chosen to be about one third of the circumferential fracture stress measured for the fibrous cap, the areas of these regions are measures for the risk of plaque fracture and may be used as a vulnerability index, i.e. a single-valued measure of plaque stability.

4.3 Variation of the Lipid Pool Stiffness

A particular feature of the computational approach over pure morphological approaches is the possibility to estimate the potential effects of drugs on plaque stability. Lipid lowering by statins reduces significantly the risk of coronary and cerebral events without noticeable plaque regression. It was hypothesized that this effect is based on lipid pool stiffening caused by a transformation of soft cholesterol esters to hard monohydrates [17].

The present mechanical analysis shows that lipid pool stiffening (and weakening) exerts only marginal effects on the stress distribution in the fibrous cap, and hence on plaque stability (Fig. 2b). As supposed in [17], the reason for this is that the surrounding tissues are much stiffer than the lipid pool, so that even a four-fold variation of its stiffness does not affect noticeably the load distribution. Hence, plaque stabilizing effects of lipid lowering therapies may be based on transformations of lipid pools to fibrous or calcified tissue, reduction of the lipid pool size, and on other mechanisms rather than on changes of the lipid composition [19,20,21]. The proposed computational approach may be helpful in evaluating the rationales of drug treatment and the underlying pathophysiological concepts.

4.4 Variation of the Media Stiffness

Another result of the present work concerning drug effects is that the media stiffness, which can be decreased by vasodilators, has a strong influence on the stress distribution in the intima. Computational analyses of this particular atherosclerotic lesion show that a softer media results in a stress increase in the fibrous cap (Fig. 2c) and a marked enlargement of the "unsafe area" (Fig. 3c), where stresses exceed the safety margin of 230 kPa.

Hence, the results based on the specific atherosclerotic lesion suggest that a decrease of the media stiffness causes higher stresses in the intima. Actually, there are many reports on ischemic syndromes triggered by various vasodilators (see, for example, [22,23]). Frequently, these syndromes were attributed to paradox vasoconstriction and steal syndromes. However, for many cases these explanations are not convincing. For the specific investigated atherosclerotic lesion the results suggest that the use of a vasodilator may lead to a higher risk of plaque disruption due to a mechanical load shift from a softened media to an unstable fibrous cap. In particular, high single doses, which occur in myocardial stress testing [23] and intoxication [22], may destabilize plaques because in these situations there is no time for adaptive remodeling. In addition, vasoconstriction is frequently observed in ruptured lesions and, thus, is suspected to be a triggering factor for plaque disruption [24]. However, the mechanical analysis shows that medial stiffening due to increased vascular tone leads to stress relief in the fibrous cap (Figs. 2c and 3c).

A conclusive implication of this study is that plaque stability can not be reduced to an intimal problem. Stress distributions are strongly influenced by non-intimal components, and hence suitable modeling of non-intimal and intimal components is equally important.

4.5   In Vivo Application

The proposed *in vitro* approach serves as a promising basis for improvements of diagnosis and treatment of unstable plaques. Yet, the visionary goal of realistic atherosclerotic lesion modeling is stability analysis for an individual patient's vessel site and subsequent determination of optimal drug treatment tailored for that particular patient. Transfer of the current *in vitro* methodology to a clinical setting requires two essential prerequisites: (i) *in vivo* plaque imaging methods, which allow distinction of major tissue components, and (ii) determination of the associated mechanical properties. Several research groups have demonstrated that MR-based plaque/wall-imaging is feasible also for *in vivo* situations [25,26,27,21], although, particularly for coronary arteries, spatial resolution and reliability of tissue characterization need to be improved for meaningful stress-strain analysis. Alternatively, invasive methods such as intravascular ultrasound [5] and yet investigational methods such as optical coherence tomography [28] could be used. Hence, there is much promise that in future morphological data of diseased vessels will be available directly from diagnostic facilities. *In vivo* determination of mechanical properties of vascular tissue components is a challenging problem, which has yet to be solved in a satisfactory way. Existing techniques are based on MR methods [29] and ultrasonic methods [30].

In summary: considering the rapid developments in bioengineering, *in vivo* plaque stability assessment based on component-specific mechanical analysis may be helpful in better characterizing patient-specific clinical situations more reliably.

4.6   Study Limitations

The present work is a method-oriented study, which does not investigate hypotheses on a statistical basis. One may argue that the results shown for one artery are highly individual and not representative for atherosclerotic lesions, in general. This is correct, however, this is also the strength of the proposed computational methodology in that it allows individual analyses. The prospect of this work is the individual diagnosis of patients with unstable plaques and the individual assessment of the potential effects of drug treatment.

For the present study an iliac artery was used, although acute ischemic syndromes following plaque fracture seem to be predominately a coronary problem. Iliac arteries, however, are more available and due to their size more convenient for preparations than coronary arteries. Basically, the proposed methodology can be also applied to coronary arteries at least for the major segments.

The proposed model considers quasi-static mechanical properties, i.e. average stress states are associated with MAP. No frequency-dependent dynamical effects were considered. Dynamical simulations require additional experimental data on the elastic frequency-dependent tissue response. *In vitro* measurements of non-diseased iliac arteries from aged human cadavers show that the incremental dynamical modulus is 30-40 % higher than the corresponding static modulus [31]. For diseased arteries, which are less distensible, the differences between dynamic and static behaviors are supposed to be even smaller, so that static solutions may approximate dynamic stress results.

Acknowledgement

We gratefully acknowledge the inspiring comments and gentle support of *Gerhard Stark*, MD, from the Division of Angiology, Medical University Graz, Austria. Financial support for this research was provided by the Austrian Science Foundation under START-Award Y74-TEC.

In Memoriam

This paper is in remembrance of our colleague and friend Christian A.J. Schulze-Bauer, MD, M.Sc., the first author of this paper, who passed away after a subarachnoid hemorrhage due to a ruptured arterio-venous malformation at December 5, 2002. He was in his 36th year of life. The investigation of the mechanical properties of human arteries was dear to his heart up to the last minute he could work.

References

1. Cheng GC, Loree HM, Kamm RD, Fishbein MC, Lee RT. Distribution of circumferential stress in ruptured and stable atherosclerotic lesions. A structural analysis with histopathological correlation. *Circ.* **87** (1993) pp. 1179-1187.
2. Loree HM, Kamm RD, Stringfellow RG, Lee RT. Effects of fibrous cap thickness on peak circumferential stress in model atherosclerotic vessels. *Circ. Res.* **71** (1992) pp. 850-858.

3. Richardson PD, Davies MJ, Born GV. Influence of plaque configuration and stress distribution on fissuring of coronary atherosclerotic plaques. *Lancet* **2** (1989) pp. 941-944.
4. Fayad ZA, Fuster V. Clinical imaging of the high-risk or vulnerable atherosclerotic plaque. *Circ. Res.* **89** (2001) pp. 305-316.
5. Hiro T, Fujii T, Yasumoto K, Murata T, Murashige A, Matsuzaki M. Detection of fibrous cap in atherosclerotic plaque by intravascular ultrasound by use of color mapping of angle-dependent echo-intensity variation. *Circ.* **103** (2001) pp. 1206-1211.
6. Humphrey JD. *Cardiovascular solid mechanics. Cells, tissues, and organs* (New York, Springer-Verlag, 2002).
7. Holzapfel GA, Stadler M, Schulze-Bauer CAJ. A layer-specific 3D model for the finite element simulation of balloon angioplasty using MR imaging and mechanical testing. *Ann. Biomed. Eng.* **30** (2002) pp. 753-767.
8. Holzapfel GA, Stadler M, Stollberger R, Regitnig P, Schulze-Bauer CAJ. A computational methodology to identify the high-risk plaque. *Ann. Biomed. Eng.*, in press.
9. Sled JG, Zijdenbos AP, Evans AC. A nonparametric method for automatic correction of intensity nonuniformity in MRI data. *IEEE Trans. Med. Imaging* **17** (1998) pp. 87-97.
10. Keeling SL. Total variation based convex filters for medical imaging. *Appl. Math. Comp*, **139** (2003) pp. 101-119.
11. Singh A, Goldgof DB, eds. *Deformable models in medical image analysis* (IEEE Computer Society Press, Los Alamitos, 1998).
12. Piegel LA, Tiller W. *The NURBS book,* 2nd ed., (New York, Springer-Verlag, 1997).
13. Holzapfel GA, Stadler M, Schulze-Bauer CAJ, *Mechanics of angioplasty: wall, balloon and stent.* In: Mechanics in Biology, Casey J, Bao G, eds., Vol. 242. 2000, The American Society of Mechanical Engineers: New York. 141-156.
14. Holzapfel GA, Gasser TC, Ogden RW. A new constitutive framework for arterial wall mechanics and a comparative study of material models. *J. Elasticity.* **61** (2000) pp. 1-48.
15. Holzapfel GA, *Nonlinear Solid Mechanics, A Continuum Approach for Engineering* (Chichester, John Wiley & Sons, 2000).
16. Schulze-Bauer CAJ, Regitnig P, Holzapfel GA. Mechanics of the human femoral adventitia including the high-pressure response. *Am. J. Physiol. Heart Circ. Physiol.* **282** (2002) pp. H2427-H2440.
17. Loree HM, Tobias BJ, Gibson LJ, Kamm RD, Small DM, Lee RT. Mechanical properties of model atherosclerotic lesion lipid pools. *Arterioscler. Thromb.* **14** (1994) pp. 230-234.
18. Cox RH. Effects of norepinephrine on mechanics of arteries in vitro. *Am. J. Physiol.* **231** (1976) pp. 420-425.

19. Helft G, Worthley SG, Fuster V, Fayad ZA, Zaman AG, Corti R, Fallon JT, Badimon JJ. Progression and regression of atherosclerotic lesions: monitoring with serial noninvasive magnetic resonance imaging. *Circ.* **105** (2002) pp. 993-998.

20. Huang H, Virmani R, Younis H, Burke AP, Kamm RD, Lee RT. The impact of calcification on the biomechanical stability of atherosclerotic plaques. *Circ.* **103** (2001) pp. 1051-1056.

21. Zhao XQ, Yuan C, Hatsukami TS, Frechette EH, Kang XJ, Maravilla KR, Brown BG. Effects of prolonged intensive lipid-lowering therapy on the characteristics of carotid atherosclerotic plaques in vivo by MRI: a case-control study. *Arterioscler. Thromb. Vasc. Biol.* 21 (2001) pp. 1623-1629.

22. Chan AW, Johnston KW, Verjee Z, Demajo WA. Delayed diagnosis of diltiazem overdose in a patient presenting with thrombosis of femoral artery. *Can. J. Cardiol.* **12** (1996) pp. 835-838.

23. Polad JE, Wilson LM. Myocardial infarction during adenosine stress test. *Heart* **87** (2002) pp. E2.

24. Bogaty P, Hackett D, Davies G, Maseri A. Vasoreactivity of the culprit lesion in unstable angina. *Circ.* **90** (1994) pp. 5-11.

25. Fayad ZA, Fuster V, Fallon JT, Jayasundera T, Worthley SG, Helft G, Aguinaldo JG, Badimon JJ, Sharma SK. Noninvasive in vivo human coronary artery lumen and wall imaging using black-blood magnetic resonance imaging. *Circ.* **102** (2000) pp. 506-510.

26. Luk-Pat GT, Gold GE, Olcott EW, Hu BS, Nishimura DG. High-resolution three-dimensional in vivo imaging of atherosclerotic plaque. *Magn. Reson. Med.* **42** (1999) pp. 762-771.

27. Toussaint JF, LaMuraglia GM, Southern JF, Fuster V, Kantor HL. Magnetic resonance images lipid, fibrous, calcified, hemorrhagic, and thrombotic components of human atherosclerosis in vivo. *Circ.* **94** (1996) pp. 932-938.

28. Jang IK, Bouma BE, Kang DH, Park SJ, Park SW, Seung KB, Choi KB, Shishkov M, Schlendorf K, Pomerantsev E, Houser SL, Aretz HT, Tearney GJ. Visualization of coronary atherosclerotic plaques in patients using optical coherence tomography: comparison with intravascular ultrasound. *J. Am. Coll. Cardiol.* **39** (2002) pp. 604-609.

29. Vinée P, Meurer B, Constantinesco A, Kohlberger B, Hauenstein KH, Petkov S. Proton magnetic resonance relaxation times and biomechanical properties of human vascular wall. In vitro study at 4 MHz. *Invest. Radiol.* 27 (1992) pp. 510-514.

30. de Korte CL, Carlier SG, Mastik F, Doyley MM, van der Steen AF, Serruys PW, Bom N. Morphological and mechanical information of coronary arteries obtained with intravascular elastography; feasibility study in vivo. *Eur. Heart J.* **23** (2002) pp. 405-413.

31. Papageorgiou GL, Jones NB. Circumferential and longitudinal viscoelasticity of human iliac arterial segments in vitro. *J. Biomed. Eng.* **10** (1988) pp. 82-90.

COMPUTATIONAL MECHANOBIOLOGY

P. J. PRENDERGAST

Centre for Bioengineering, Department of Mechanical Engineering, Trinity College, Dublin, Ireland
E-mail: pprender@tcd.ie

Mechanical forces act at all levels in the biological system, from the organ level down to individual molecules. From observations and experiments, it has been found that these forces play a role in the health of the tissues. The question to ask is: How? By what mechanism are mechanical forces able to regulate tissue biology? The answer involves finding out how forces on the whole organ regulate gene expression in cells, or cause cells to change to a new cell type (i.e. differentiate). The relationship between mechanical forces and tissue biology has been investigated using computational modeling, and this paper begins with a review of that work. Next, an attempt is made to present a methodology for simulation of mechano-regulation. This requires the derivation of equations that relate the tissue state at time t_0 to the tissue state at $t > t_0$ when the mechanical loading is known. The final step is confirmation of these equations against observations made during specific processes known to be mechano-regulated; the two examples chosen are fracture healing and osteochondral defect healing. If computational tools capable of simulating tissue responses to stress can be derived, then it becomes possible to simulate the consequences of mechanical environment on growth, adaptation, and ageing.

1 Introduction

Mechanical forces constantly act on tissues and organs. Sometimes the consequences of mechanical forces are obvious, such as when a bone fractures or a ligament is injured by over-stretching. Ascenzi [1] describes some of Galileo's researches on the strength of bones. From 1938 to 1963, a comprehensive study of all tissues was performed by the Japanese anatomist Hiroshi Yamada and reported in *The Strength of Biological Tissues* [2]. A surge of research is now ongoing to accurately determine the mechanical properties of tissues. Comprehensive information can be found for bone, cartilage, ligament, tendon, and vascular tissues [3,4,5,6].

Knowledge of how tissues deform and fail is important, but it is arguable that it is even more important to know how mechanical forces act on tissues to maintain health and to regulate biological processes. This aspect of biomechanics has recently been given the name 'mechanobiology'. In *Skeletal Function and Form* [7] many examples are given of mechanobiology of skeletal development, ageing, and regeneration. Van der Meulen & Huiskes [8] have presented a review of the literature of both experimental and computational studies in mechanobiology. As they put it, the premise of mechanobiology is that biological processes are regulated by signals to cells generated by mechanical loading, and they go on to describe experimental mechanobiology investigations using living tissues: e.g. a fractured long bone can be exposed to various degrees of bending and the corresponding tissue formation observed. Computational mechanobiology aims to discover how mechanical forces acting on the organ (e.g., a whole bone) translate into stimuli acting on the cell so as to regulate tissue growth, remodeling, adaptation, and ageing (Fig. 1).

Focusing on computational mechanobiology, we can see that computational mechanobiology problems may be split into two parts:

1. Solution of problem to determine cell deformations when the forces on the organ (e.g., whole bone) are known. This is represented schematically in the lower part of figure 1. This involves determination of constitutive models of the tissues so as organ-to-tissue computations can be performed, and it requires micro-mechanical modeling to perform the tissue-to-cell computation.

2. Derivation of equations to relate the mechanical stimuli to adaptation of the tissue/organ construct. This involves modeling of the feedback loops presented in the upper part of figure 1.

It is the second one of these aspects of computational mechanobiology that is considered in this Chapter.

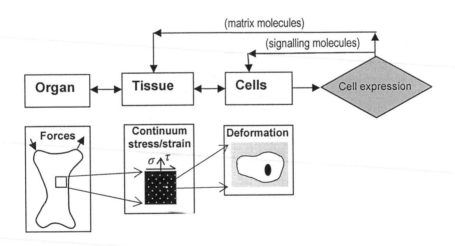

Figure 1. Organs are subjected to mechanical forces, forces create stresses and strains within the tissue matrix, and strains cause cell deformation. Cell deformations can modulate all manner of cell responses, including the expression of signalling molecules and matrix-forming proteins. They may also stimulate the cell to differentiate.

2 A Review of Computational Mechanobiology

Growth, adaptation, and remodeling of tissues are caused by processes active inside the tissue. These processes are executed by cells. They involve changing the tissue from one phenotype to another, or by replacing it altogether; either way a new cell population becomes resident at the place where differentiation has occurred. In the mathematical sense, we can consider a domain filled with biological tissue as a reactive continuum.

One cell type that is central to this process is the *stem cell*, the parent cell of cells that generate connective tissues. According to Bianco & Robey [9], the concept of a stem cell originated in the early 20th Century. Recently researchers have attempted to elucidate what controls stem cell differentiation. It has been suggested that chemical and mechanical stimuli can control the differentiation of adult skeletal stem cells (also called mesenchymal stem cells) into either fibrous connective tissue, cartilaginous tissue, bone, or adipose tissue [10]. Many researchers have attempted to establish the relationship between mechanical stressing on undifferentiated tissue and the ultimate tissue phenotype formed. The modern treatment of the subject, however, begins with Pauwels.

2.1 Pauwels' Theory of Tissue Differentiation

In the author's opinion, Pauwels' theory about tissue differentiation can be conveniently divided into two hypotheses. The first hypothesis makes the statement that alterations in mesemchymal stem cell shape effect mesenchymal stem cell differentiation so that tissues are formed as follows:

Hypothesis 1
Pure alteration in cell shape stimulates synthesis of collagenous fibrils whereas pure alteration in cell volume stimulates synthesis of cartilaginous tissues.

The second hypothesis makes a statement about the control of ossification (ossification is bone formation) and suggests that the soft tissues may be necessary to reduce mechanical stresses to allow ossification. It may be given as follows:

Hypothesis 2
"The construction [of bone] cannot be carried out immediately with the final element ... because the blasteme cell cannot differentiate into bony tissue where mechanical stressing directly acts on it". (Pauwels [11])

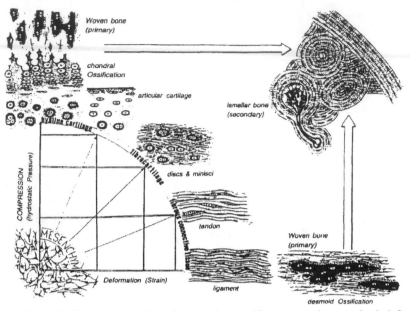

Figure 2. Pauwels' diagram encapsulating his hypotheses with respect to how mechanical forces regulate tissue differentiation and ossification.

Pauwels [12] combined these two hypotheses diagrammatically to illustrate his idea that the formation of either fibrous connective tissue or cartilaginous tissue was required to stabilize the mechanical environment before ossification could occur. The diagram shown in figure 2 is taken from Weinans and Prendergast [13]. It is Hypothesis 1 that informs what appears on the x axis (deformation by shear strain) and the y axis (hydrostatic pressure). The larger arrows indicate that, as time passes, ossification of these soft tissues occurs, but may be inhibited, or at least slowed down, according to Hypothesis 2.

2.1.1   Hypothesis 1: The relationship between cell deformation and differentiation

Pauwels tested *Hypothesis 1* by analyzing the stresses generated in tissues during fracture healing. Using simple physical models, he showed that the tissue types within a healing fracture callus could be correlated with the hydrostatic or deviatoric nature of the stress in the regenerating tissue. This was further confirmed by the finite element analyses of Carter and co-workers [7], and not just for fracture healing [14,7,15], but for other problems such as distraction osteogenesis and chondrogenesis during bone growth [16]. Carter and co-workers extended *Hypothesis 1* by specifying how a lack of oxygen would inhibit osteogenesis and maintain the chondrogenic phenotype. More recently Claes and Heigele [17] and Gardner *et al.* [18] have attempted to quantify the nature of the stress by analyzing a real callus. In both cases the mechanical environment occurring in cartilage about to ossify could be identified; Claes and Heigele [17] found that endochondral ossification was predicted if the absolute value of the strain was between 5% and 15%, and the compressive pressure lower than -0.15 MPa.

The hypothesis introduced by Prendergast *et al.* [19] is that substrate strain and fluid flow are the primary biophysical stimuli for stem cell differentiation. This hypothesis is proposed based on empirical evidence that differentiated cells respond to biophysical stimuli, e.g., Kaspar *et al.* [20] placed osteoblasts on a plate subjected to four-point bending and recorded increases in matrix synthesis (collagen Type I) and reductions in the synthesis of certain signaling molecules; in a similar experiment, Owan *et al.* [21] found that fluid flow was a dominant stimulus over substrate strain. Klein-Nulend *et al.* [22] analyzed a layer of osteocytes, osteoblasts, and fibroblasts and found that osteocytes generated increased signaling molecules (PGE$_2$) in response to pulsatile fluid flow and decreased in response to hydrostatic pressure. Prendergast *et al.* [19] used a poroelastic theory to compute the fluid flow and substrate strain in the tissue, and these are used as a basis for the stimulus for cell differentiation as follows:

$$S = \frac{\gamma}{a} + \frac{v}{b} \tag{1}$$

where γ is the shear strain and v is the fluid velocity, a and b being empirical constants.

2.1.2   Hypothesis 2: The route to ossification

Turning to *Hypothesis 2*, it seems that Pauwels made no direct test of this hypothesis himself. In any case it is difficult to test because it requires that the mechanical environment is controlled after the soft-issue is formed. Recent experiments (e.g., Aspenberg *et al.* [23]) have attempted to do this. From a computational mechanobiology point-of-view, it requires the inclusion of time as a variable and so simulations, rather than analyses at discrete times, are required.

Some of the most influential papers describing how to simulate adaptation of biological tissue to a change in load were by Cowin and co-workers. Beginning with Cowin & Hegedus [24], an adaptive elasticity theory was proposed whereby a unit of bone tissue was modeled as an open system, and an equation for the rate of change of density as a function of the strain was developed. Adaptive elasticity, and the numerical implementations of it, did not rely explicitly on knowledge of how cells responded to strain in the way proposed by Pauwels. They approached the problem according to continuum mechanics and attempted to relate the rate of change of mass to various continuum variables. The continuum approach was often used over the next two decades (see [25, 26]). The results of Weinans *et al.* [27]

suggested a rethink of the continuum approach because an instability in local remodeling could generate a pattern analogous to a trabecular bone network. Although the instability was attributed to the discretization of the finite element mesh and was therefore inadmissible, Jacobs *et al.* [28] later showed that a similar process occurred even if these numerical problems were corrected. Huiskes *et al.* [29] showed how explicit modeling of cell behavior could be used to predict trabecular bone remodeling in response to changes in the applied load.

Simulation of the time-course of tissue differentiation in the way proposed by *Hypothesis 2* of Pauwels was presented by Huiskes *et al.* [30] using the stimuli of strain in the solid matrix and fluid flow (as proposed in [19]) for tissue differentiation in the tissue surrounding an implant. Using experimentally-measured stiffnesses for the differentiating tissues, they computed that micromotion would reduce only if the appearance of the soft tissues would eventually prevent the implant micromotion so as ossification could occur. An attempt to illustrate this is given in figure 3 below. The solid line shows what would occur in an environment where a high shear persists (i.e. maintenance of fibrous tissue and inhibition of ossification) whereas the dashed line shows what would occur *if the presence of the soft tissue could progressively reduce the micromotions* (i.e. ossification would occur). The 'route to ossification' is only opened if the soft-tissue progressively reduces mechanical stressing of the mesenchymal stem cells [31].

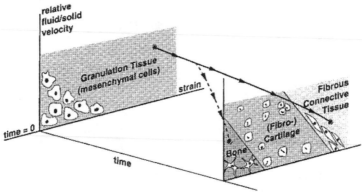

Figure 3. The back plane shows that, as regeneration begins in the granulation tissue in which some strain/fluid velocity stimulus exists. As time passes, the stimulus at all locations will change as cells begin to synthesis matrix molecules. If the creation of matrix does not progressively cause a reduction in the magnitude of the stimulus (e.g. under a motion-control situation), then soft-tissues persist and ossification is inhibited. If the presence of the soft-tissues progressively reduces the stimulus, then a mechanical environment of low stimulus will eventually appear (dashed line). In this new environment, ossification can occur. This is ossification path shown by the large arrows in Pauwels diagram in figure 2. Taken from Prendergast et al. [19].

The aim of the following sections is to further develop these ideas, and to explore the degree to which the concepts of Pauwels' and others, can be set up for computational solutions. In particular, the challenge of better accounting for the behaviour of the various cell populations in computational mechanobiology solutions still needs to be met. If this were possible, then the experiments being performed to determine cell responses to various stimuli could be used to "empirically inform" computational mechanobiology research.

3   Evolution Equations

When considering a mathematical framework for simulating mechano-regulated biological processes, it can be noted that the processes are thermodynamically irreversible (Weinans & Prendergast [13]). In this respect, it is appropriate to consider equations of evolution which describe in precise terms the way these processes evolve. And to quote Fung [32],

> "….some new hypotheses must be introduced, whose justification can only be sought by comparing any theoretical deductions with experiments".

Stem cells differentiate into the cells which produce the various tissue phenotypes of interest (*i.e.*, the connective tissues, which are fibrous connective tissue, fibrocartilage, cartilage, and bone). Therefore we must compute a distribution of cells throughout the continuum. If n^i denotes the number of cells of the ith type then,

$$\frac{dn^i}{dt} = D^i \nabla^2 n^i + P^i(S).n^i - K^i(S).n^i, \qquad (2)$$

where D^i is a diffusion coefficient for cell i, $P^i(S)$ is a proliferation rate and $K^i(S)$ is an apoptosis (death) rate for cell i as a function of the stimulus S. If ϕ_j denotes the volume fraction of a tissue type j then

$$\sum_{j=1}^{n_t} \phi_j = 1, \qquad (3)$$

where n_t is the number of tissue types. The diffusion coefficient of cells of type i through a volume of tissue can be approximated as the weighted average of the tissue types present in the volume, i.e.

$$D^i = \sum_{j=1}^{n_t} D_{ij} \phi_j, \qquad (4)$$

D_{ij} being the diffusion rate of cell i in tissue j. The proliferation rate may be independent of the stimulus or, more generally, an optimal stimulation for proliferation may exist requiring a relationship of the form:

$$P^i(S) = a_i + b_i S + c_i S^2. \qquad (5)$$

Whether or not a stem cell will differentiate into a fibroblast to form fibrous connective tissue, a chondrocyte to form cartilage, or an osteoblast to form bone, is hypothesized to depend on the stimulus, S. This is a central tenant of mechanobiology, promoted by the concepts of Caplan [10] and assumed by researchers in the field; see Prendergast & Van der Meulen [33]. If the stimulus based on strain and flow is used (Eqn.(1)), then it can be written as:

$$
\begin{array}{ll}
0 \leq S < n & \text{bone resorption,} \\
n \leq S < 1 & \text{bone,} \\
1 \leq S < m & \text{cartilage,} \\
m \leq S & \text{fibrous connective tissue,}
\end{array}
\qquad (6)
$$

where the stimulus S is a function of the biophysical stimuli on the cells that will change the cell shape or otherwise stimulate the cell. It is very difficult if impossible to determine what these stimuli may be *a priori*; instead some new hypothesis must be introduced and its validity assessed by comparing predictions with observations.

To derive an evolution equation, the magnitude of the stimulus needs to be related to the rate of tissue formation. If it is assumed that the rate of change of tissue density ρ is a function of the difference between the actual stimulus and the homeostatic stimulus ΔS then:

$$\frac{d\rho_i}{dt} = f(\Delta S, n_i) \qquad (7)$$

Detailed forms of Eqn. (7) have been proposed for bone, that is for $0 = S < n$, (the bone resorption field) and $n=S<1$ (the bone formation field). In particular, the rate of bone formation and the rate of bone resorption have been described by Carter [14] as non-linear functions of strain with a "zone" close to homeostatic equilibrium, or a "lazy zone", i.e.

$$\frac{d\rho}{dt} = \begin{cases} f_1(S) & S < S_{min} \\ 0 & S_{min} < S < S_{max} \\ f_2(S) & S > S_{max} \end{cases} \qquad (8)$$

In Prendergast [34], the specific formulation of Eqn. (8) was developed to include microdamage as a stimulus. Using the concept that damage accumulation is a stimulus for bone remodeling [35], it can be noted that, if the repair rate in bone is not fast enough to repair damage as it forms, then damage will accumulate. This net damage will reduce the Young's modulus of the matrix surrounding a cell and cause a reduction in the stress sensed by the cells leading to resorption. Therefore, because of the influence of damage, bone resorption occurs at high strain levels. This will result in an evolution equation of the form plotted in figure 4 below.

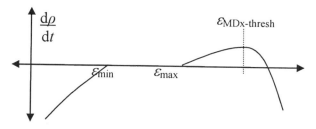

Figure 4. The rate of change of bone density as a function of the stimulus (the stimulus is strain only in this case). At low strain resorption occurs, at intermediate strains there is a zone of equilibrium, at higher strains there is deposition of new bone in an attempt to reduce the stresses, and above some threshold there is damage accumulation to the extent that the rate of deposition of bone reduces. If the stress is too high, pathological resorption occurs.

If this concept can be extended to all connective tissues, then a superimposed set of evolution equations could be developed to predict the evolution of all tissue phenotypes under the influence of the stimulus as shown in figure 5. The set of evolution equations represented in figure 5 will ensure resorption of all connective tissue when $S<S_a$, maintenance of bone and resorption of cartilage and fibrous connective tissue if $S_a<S<S_b$, formation of bone at an ever increasing rate, and resorption of cartilage and fibrous connective tissue if $S_b<S<S_c$, damage accumulation in bone and a reduced bone formation rate, maintenance of cartilage, and resorption of fibrous connective tissue if $S_c<S<S_d$, resorption of bone and fibrous connective tissue and formation of cartilage at an ever increasing rate for $S_d<S<S_e$, and so on. It must be stated that the set of evolution equations represented by figure 5 is highly speculative and serves mainly to represent the nature of the problems awaiting solution in computational mechanobiology.

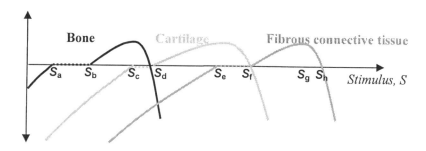

Figure 5. A representation for hypothesized set of equations for describing the rate of formation of various connective tissues as a function of mechanical stimulus (see text for meaning of the S variables shown on the x-axis).

4   Computational Implementation

In the case of regeneration of bone, we may let i take the values as follows: 1= stem cells (precursor cells), 2 = fibroblasts, 3 = chondrocytes, and 4 = osteoblasts. The author's research group have considered two analyses of bone regeneration; long bone fracture repair [36,37,38]) and osteochondral defect healing [39].

4.1   Fracture repair

When a long bone such as the tibia is fractured an inflammation phase immediately begins and granulation tissue containing mesenchymal stem cells originating from the medullary cavity, the external muscle, or the surface of the bone (cadmium layer of the periostium), enters the fracture callus. Various tissues are generated in an orchestrated sequence leading to eventual healing of the bone. To simulate this process according to the formulation described in Section 3 above, Eqn. (2) was simplified to

$$\frac{dn}{dt} = D \ \nabla^2 n \qquad (9)$$

such that proliferation and apoptosis were neglected. It was assumed that the only cell that differentiated was the stem cell. In these simulations, the constants were

$$a = 3.75\%$$
$$b = 3.0\mu ms^{-1}$$
$$n = 0.01 \qquad (10)$$
$$m = 3.0$$

and the diffusion constant was taken as $D = 0.34$ mm^2 per day. An asymmetric finite element model of a fracture callus was generated, and a poroelastic model of the tissues were used to calculate apparent levels of fluid flow and tissue strains. Equation 6, with the constants as given in Eqn. (10), was used to determine which tissue will form at a site (element) in the model. The evolution equations had the character of an algorithm whereby if S exists at a site (i.e. in an element) for more than some transformation time, then tissue j is generated such that $\phi_{i=j} = 1$ and

$\phi_{i \neq j} = 0$.   Clearly   this   kind   of   evolution   equation   assumes   immediate transformation of the tissue at a site, which is only good as a first approximation. This computational scheme was set up in an iterative procedure. Beginning with a fracture callus formed from granulation tissue post-inflammation phase, the sequence of tissue formations and resorption leading to a healed bone was simulated (Fig.6) in both 2D [36,37], and in 3D [38].

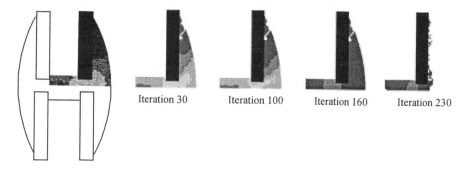

Iteration 30   Iteration 100   Iteration 160   Iteration 230

Figure 6. Computer simulation of the process of fracture healing of a long bone under 500 N loads with a fracture gap of 6 mm. Diagram adapted from Lacroix and Prendergast [36].

4.2 Osteochondral Defect Repair

Occasionally, the cartilage layer on a bone surface becomes torn or damaged and in need of repair. The repair procedure can involve a surgical technique whereby the subchondral bone below the defect is drilled, a cell-seeded scaffold is placed in the defect, and it is expected that the new cartilagetissue will eventually form in the defect. Kelly [40] reports an attempt to simulate this process by considering cell proliferation into the defect filled with granulation tissue, followed by dispersion of cells according to Eqn. (2) whereby proliferation, differentiation, and apoptosis is a function of the stimulus. The proliferation equation of Eqn. (4) is written as:

$$\frac{d}{dt}\begin{Bmatrix} n^1 \\ n^2 \\ n^3 \\ n^4 \end{Bmatrix} = \begin{bmatrix} a_1 & b_1 & c_1 \\ a_2 & b_2 & c_2 \\ a_3 & b_3 & c_3 \\ a_4 & b_4 & c_4 \end{bmatrix} \begin{Bmatrix} 1 \\ S \\ S^2 \end{Bmatrix} \tag{11}$$

where

$$\begin{bmatrix} a_1 & b_1 & c_1 \\ a_2 & b_2 & c_2 \\ a_3 & b_3 & c_3 \\ a_4 & b_4 & c_4 \end{bmatrix} = \begin{bmatrix} a_{\text{stem\_cell}} & 0 & 0 \\ a_{\text{fibroblast}} & b_{\text{fibroblast}} & c_{\text{fibroblast}} \\ a_{\text{chondrocyte}} & 0 & 0 \\ a_{\text{osteoblast}} & 0 & 0 \end{bmatrix} \tag{12}$$

The algorithm for the solution of the full tissue differentiation model is shown in figure 7. These simulations successfully predict the formation of fibrous connective tissue, cartilage and bone within the defect. The simulations predict that fibrous tissue persists at the surface of the defect and full healing of the cartilage layer does not ensue. This is similar to what is found clinically. Simulations have also been done with various amounts of mechanical properties of scaffold and it was found that an optimal scaffold geometry may exist.

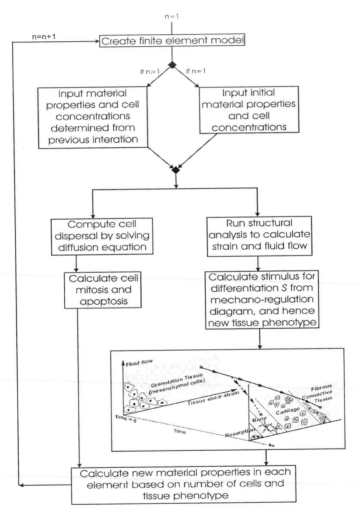

Figure 7. An algorithm for simulation of tissue differentiation, including cell differentiation in response to biophysical stimuli, and cell migration and proliferation. From Kelly [40].

5 Conclusions and Outlook

The maintenance of life in a gravitational field relies on the fact that living organisms have evolved biological materials that can resist mechanical forces. These biological materials have, under the selection pressures of evolution, combined to form structures – musculo-skeletal structures controlled by the central nervous system. These structures are capable of gathering food, surviving, reproducing and, in the end, continuing the species. The regulation of tissue biology in response to mechanical forces plays a central role in this achievement. For this reason, questions about how tissues are mechano-regulated are of fundamental importance across a range of disciplines. This paper has attempted to show how we, and others, have attempted to use computational mechanics to address these questions.

Besides being of interest for fundamental reasons, computational mechanobiology can also play a role in the design of load-bearing implants, such as orthopedic and cardiovascular devices. These implants alter the stress distribution in the tissue around them, and prediction of the reaction of the tissues to the stress change is of central importance in implant design and optimization [7,41,42].

Clearly the above scheme for simulation of mechano-regulation of tissues contains a series of hypotheses about how cells react to physical stimuli: Eqn. (2) assumes that the cells disperse according to a diffusion equation which may not be true if the cells are convected into the regenerating region by blood flow or if they crawl in response to chemoattractants [43,44]. Eqn. (2) also simplifies the stimulus to mitosis (cell division = proliferation) to a mechanical stimulus S and this is certainly a simplification. Eqn. (5) and Eqn. (6) suggest that the same stimulus mechano-regulates mitosis as regulates stem cell differentiation and a more complex relationship is certain to exist in reality.

Of the issues that warrant further research, one of the most interesting is the nature of the biophysical stimulus. Carter and Beaupré [7] propose hydrostatic and deviatoric stresses in an elastic model of the tissue and Kuiper et al. [45] derive fluid shear stresses as a stimulus. Either of these may be superior to the one presented here. Furthermore, tensorial representations of the stimulus (e.g. Doblaré et al., [46]) give a better representation of the complexity of tissue adaptation and remodeling. Micromechanical finite element models which allow the computation of the stimulus as a function of the local microstructure are needed if the results on cell experiments are to be used to predict changes in apparent-level properties. In the author's opinion this is one of the most pressing problems to be solved in computational mechanobiology.

The relationship between mechanical stimulus in a tissue and the rate of formation or resorption presented in figure 5 is quite speculative. The only part that has been explored with rigor to date is $S<S_c$, i.e., the part dealing with the resorption, homeostasis, and deposition of bone. However, if these methods are to find use in the new discipline of tissue engineering, the other parts of this

relationship will need to be developed. This will require close collaboration between cell and molecular biologists, applied mechanicians, and tissue histologists in a broad interdisciplinary collaboration.

Acknowledgments

Financial support for this research is provided by the Programme for Research in Third Level Institutions (PRTLI) of the Higher Education Authority in Ireland. It is a pleasure to acknowledge the scientific contributions of Dr Damien Lacroix, (now with the Department of Material Science and Metallurgical Engineering, Technical University of Catalonia, Spain) and Dr Danny Kelly (now with Clearstream Technologies Ltd, Ireland).

References

1. Ascnezi A., Biomechanics and Galilei Galileo. *J. Biomech.* **26** (1993) pp. 95-99.
2. Yamada H., *Strength of Biological Materials.* (Williams 'and Wilkins Co., Baltimore, 1970).
3. Black J., Hastings G. Eds., *Handbook of Biomaterial Propertie.* (Chapman and Hall, London 1998).
4. Cowin S.C., *Bone Mechanics Handbook.* Cowin S.S. Ed. (CRC Press, Boca Raton 1999).
5. Humphrey J., *Cardiovascular Solid Mechanics: Cells, Tissues, Organs.* (Springer, Berlin 2002).
6. Mow V.C., Kuei S.C., Lai W.M., Armstrong C.G., Biphasic creep and stress relaxation of articular cartilage: theory and experiments. *J. Biomech. Eng.* **102** (1980) pp. 73-84.
7. Carter D.R., Beaupré G.S., *Skeletal Function and Form. Mechanobiology of Skeletal Development, Aging, and Regeneration.* (Cambridge University Press, Cambridge 2001).
8. Van der Meulen M.C.H., Huiskes R., Why mechanobiology? A survey article. *J. Biomech.* **35** (2002) pp. 401-414.
9. Bianco P., Robey P.G., Stem cells in tissue engineering. *Nature* **414** (2001) pp. 118-212.
10. Caplan A.I., The mesengenic process. *Clin. Plast. Surg.* **21** (1994) pp. 429-435.
11. Pauwels F., *Atlas zur Biomechanik der Gesunden und Kranken Hüffe.* (Springer, Berlin 1973).
12. Pauwels F., *Biomechanics of the Locomotor Apparatus.* (Springer, Berlin 1980) p. 118.

13. Weinans H., Prendergast P.J., Tissue adaptation as a dynamical process far from equilibrium. *Bone* **19** (1996) pp. 143-149.
14. Carter D.R., Mechanical loading histories and cortical bone remodeling. *Calcified Tissue Int.* **36** (1984) pp. 19-24.
15. Carter D.R., Blenman P.R., Beaupré G.S., Correlations between mechanical stress history and tissue differentiation in initial fracture healing. *J. Orthop. Res.* **6** (1988) pp. 736-748.
16. Carter D.R., Beaupré G.S., Giori N.J., Helms J.A., Mechanobiology of skeletal regeneration. *Clin. Orthop.* **355S** (1998) pp. 41-55.
17. Claes L.E., Heigele C.A., Magnitudes of local stress and strain along bony surfaces predict the course and type of fracture healing. *J. Biomech.* **32** (1999) pp. 255-266.
18. Gardner T.N., Stoll T., Marks L., Mishra S, Tate M.K., The influence of mechanical stimulus on the pattern of tissue differentiation in a long bone fracture – an FEM study. *J. Biomech.* **33** (2000) pp. 415-425.
19. Prendergast P.J., Huiskes R., Søballe K., Biophysical stimuli on cells during tissue differentiation at implant interfaces. *J. Biomech.* **30** (1997) pp. 539-548.
20. Kaspar D., Seidl W., Neidlinger-Wilke C., Ignatius A., Claes L. Dynamic cell stretching increases human osteoblast proliferation and CICP synthesis but decreases osteocalcin synthesis and alkaline phosphatase activity. *J. Biomech.* **33** (2000) pp. 45-51.
21. Owan I., Burr D.B., Turner C.H., Qui J., Tu Y., Onyia J.E., Duncan R.L., Mechanotransduction in bone: osteoblasts are more sensitive to fluid forces than mechanical strain, *Am. J. Appl. Physiol.* **273** (Cell Physiology 42) (1997) pp. C810-815.
22. Klein-Nulend J., Semeins C.M., Burger E.H., van der Plas A., Ajubi N.E., Nijweide P.J., Response of isolated osteocytes to mechanical loading in vitro, In *Bone Structure and Remodelling*, A. Odgaard and H. Weinans Eds. (World Scientific, Singapore 1996) pp. 37-49.
23. Aspenberg P., Goodman S., Toksvig-Larsen S., Ryd L., Albrektsson T., Intermittent micromotion inhibits bone ingrowth in Titanium implants in rabbits. *Acta. Orthop. Scand.* **63** (1992) pp. 141-145.
24. Cowin S.C., Hegedus D.H., Bone remodeling I: theory of adaptive elasticity, *J. Elasticity* **6** (1976) pp. 313-326.
25. Huiskes R., Chao E.Y.S., A survey of finite element method in orthopaedic biomechanics – the first decade. *J. Biomech.* **16** (1983) pp. 385-409.
26. Prendergast P.J., Finite element modeling in tissue mechanics and orthopaedic implant design. *Clin. Biomech.* **12** (1997) pp. 343-366.
27. Weinans H., Huiskes R., Grootenboer H.J., The behaviour of adaptive bone remodelling simulation models, *J. Biomech.* **25** (1994) pp. 1425-1441.

28. Jacobs C.R., Levenston M.E., Beaupré G.S., Simo J.C., Carter D.R., Numerical instabilities in bone remodelling simulations: the advantages of a node-based finite element approach, *J. Biomech.* **28** (1995) pp. 449-459.
29. Huiskes R., Ruimerman R., van Lenthe G.H., Janssen J.D., Effects of mechanical forces on maintenance and adaptation of form in trabecular bone. *Nature* **405** (2000) pp. 704-706.
30. Huiskes R., van Driel W.D., Prendergast P.J., Søballe K., A biomechanical regulatory model of periprosthetic tissue differentiation. *J. Mat. Sci.: Mater. Med.* **5** (1997) pp. 785-788.
31. Prendergast P.J., Huiskes R., An investigation of Pauwels' mechanism of tissue differentiation. *In Proceedings of the Fifth Conference of the European Orthopaedic Research Society* (Munchen, Germany 1995) p. 76.
32. Fung Y.C. *Foundations of Solid Mechanics*. (Prentice-Hall, Englewood Cliffs 1965) p. 360.
33. Prendergast P.J., van der Meulen M.C.H., Mechanics of bone regeneration, *In Bone Mechanics Handbook*, S.C. Cowin Ed. (CRC Press, Boca Raton 2001) pp. 25-1 to 25-18.
34. Prendergast P.J., Mechanics applied to skeletal ontogeny and phylogeny. *Meccanica* **37** (2002) pp. 317-334.
35. Prendergast P.J., Taylor D., Prediction of bone adaptation using damage accumulation, *J. Biomech.* **29** (1994) pp. 1067-1076.
36. Lacroix D., Prendergast P.J., A mechano-regulation model for tissue differentiation during fracture healing: analysis of gap size and loading, *J. Biomech.* **35** (2002) pp. 1163-1171.
37. Lacroix D., Prendergast P.J., Three-dimensional simulation of fracture repair in the human tibia, *Comp. Meth. Biomech. Biomed. Eng.* **5** (2002) pp. 369-376.
38. Lacroix D., Prendergast P.J., Li G., Marsh D., Biomechanical model to simulate tissue differentiation and bone regeneration: application to fracture healing. *Med. Biol. Eng. & Comput.* **40** (2002) pp. 14-21.
39. Kelly D.J., Prendergast P.J., The poor quality of repair tissue that forms in osteochondral defects can be attributed to the local mechanical environment within the defect, *In Proceedings of the 9th Bioengineering Conference,* D.P. FitzPatrick et al. Eds. (Ireland 2003) p. 75.
40. Kelly D.J., *Mechanobiology of Tissue Differentiation During Osteochondral Defect Repair*. PhD Thesis, University of Dublin (2003).
41. Prendergast P.J., Contro R., *Meccanica (Special Issue)* **37** (2002) No. 4/5.
42. Van der Meulen M.C.H, Prendergast P.J., Mechanics in skeletal development, adaptation and disease. *Phil. Trans. R. Soc. (Lond.)* **358** (2000) pp. 565-578.
43. Bailon-Plaza A., Van der Meulen M.C.H., A mathematical framework to study the effects of growth factor influences on fracture healing. *J. Theor. Biol.* **212** (2001) pp. 191-209.

44. Murray J.D., *Mathematical Biology.* (Springer, Berlin, 2nd Edn. 1993) pp. 232-236.

45. Kuiper J.H., Ashton B.A., Richardson J.B., Computer simulation of fracture callus formation and stiffness restoration. *In: Proceedings of the European Society of Biomechanics,* P.J. Prendergast, T.C. Lee, A.J. Carr, Eds. (Royal Academy of Medicine in Ireland, Dublin 2000) p. 61(www.biomechanics.ie/esb2000)

46. Doblaré M., Garciá J.M., Cegoñino J., Development of an internal bone remodelling theory and applications to some problems in orthopaedic biomechanics, *Meccanica* **37** (2002) pp. 365-374.

OPTIMIZATION MODELS IN THE SIMULATION OF THE BONE ADAPTATION PROCESS

H. RODRIGUES AND P. R. FERNANDES

IDMEC – Instituto Superior Técnico, Lisboa,
Portugal
E-mail: hcr@ist.utl.pt, prfernan@dem.ist.utl.pt

This chapter describes material optimization models and their application to the simulation of the bone adaptation process due to mechanical loading.

In 1892 Julius Wolff [1] described the dependency of bone on applied loads in his "Law of Bone Remodelling": The internal bone architecture depends on stresses and the trabeculae are aligned with principal stress directions. Following Wolff observations this chapter presents models for bone adaptation/remodelling based on material optimization models. Bone is assumed as a cellular material with variable relative density. The changes on bone density and orientation are obtained by the minimization of a cost function accounting for both the structural stiffness and the biological cost associated with metabolic maintenance of the bone tissue. The effective bone properties are obtained by an asymptotic homogenization method. These homogenized properties are computed for the cellular material obtained by the period repetition of a given base cell, thus with a given material symmetry, or alternatively solving a local material optimization problem in order to introduce the material symmetry as a variable The models described are analyzed and their relations and analogies with other models presented in the literature are discussed.

1   Introduction

Since Wolff originally proposed that the adaptation of trabecular bone to its mechanical environment could be described by mathematical rules (Wolff [1], Treharne [2]) researchers have developed progressively more sophisticated and complete mathematical models to predict this behaviour. With the advent of modern techniques in computational mechanics these models have been implemented numerically and have proved their importance not only as a tool to predict adaptation resulting from orthopaedic interventions but also to further our understanding of the characteristics of the underlying biological mechanisms.

In the majority of such models, trabecular bone has been described only in terms of its apparent or relative density. An empirical local equation is then applied at each point in the tissue for the rate-of-change of relative density (resorption or apposition) in terms of the current relative density and the local mechanical strain/stress tensor. Isotropic material behaviour is assumed with an experimental power law relationship between Young's Modulus and relative density. Formulations of this type have been quite successful in predicting a wide variety of both naturally occurring bone morphologies (see e.g. Fyhrie and Carter [3], Weinans et al. [4], Mullender et al. [5]) as well as simulating bone adaptation subsequent to orthopaedic procedures. Despite their successes, a significant shortcoming has been their treatment of trabecular bone as an isotropic material and thus disregarding orientation in the remodelling equations.

Experimental tests on the mechanical properties of trabecular bone indicate that its material properties range from having little or no dependence on direction (isotropy) to being strongly dependent on direction (anisotropy) (Brown and Fergurson [6], Yang et al. [7], Odgaard et al. [8]). This has led to a growing attention on the issue of more general bone adaptation formulations not based on material isotropy and thus coupling material density and orientation into a single model (see e.g. Jacobs et al. [9], Hollister et al. [10] Garcia et al. [11] and Doblaré and Garcia [12]). One approach to obtain such coupling is to model bone with a less restrictive class of elastic symmetry, namely assuming it as a cellular material with an orthotropic microstructure and identifying remodelling with a material optimization process (see e.g. Fernandes et. al. [13]). A formulation of this type has the advantage that the material now has an orientation which may vary along the bone domain. It can be shown in general that the optimal solution is achieved when the microstructure is locally aligned with the local principal strain directions as observed in Wolff's Law.

This strategy is sufficient when only a single load case is considered. In the case of multiple loads the optimal orientations associated with each load case need not coincide, making an optimal orthotropic microstructure problematic. In this case it is expected that an optimal microstructure would no longer be orthogonal (see Pidaparti and Turner, [14] in the case of trabecular bone). One approach to account for this situation was proposed by Jacobs et al. [9]. Essentially the method consists on finding the full anisotropic elasticity tensor which is optimal for the particular strains at each point in the bone. This problem can be solved analytically, and as such no modelling of the microstructure is required. Even though this approach is more general than those that depend on an a priori assumption of orthogonal symmetry, it leads to a question of realizability. It was shown that in some cases the optimal anisotropic elasticity tensor is strongly directional and did not fit well the experimental data. In these cases the optimal elasticity tensor may exceed what bone is able to achieve, and indeed may not correspond to any two phase cellular microstructure at all.

To overcome this difficulty Rodrigues et al. [15] presented a model for bone adaptation which avoids any a priori assumptions of material symmetries while restricting the material behaviour to achievable two-phase microstructures. The model adopts a global-local hierarchical approach in which a global model of an entire bone supplies strain and density information to a series of local models of optimal microstructure at each material point of the global model. These microstructural models, in turn, provide the global model with an optimal two-phase microstructure and the corresponding elasticity tensor. In order to maintain a reasonable complexity of the microstructure, this is generated by an optimal design technique working at a single design scale (thus multi-scale ranked composites are excluded).

2   Material Model for Trabecular Bone

2.1   The bone microstructure

Trabecular bone is a cellular material, non-homogeneous and anisotropic (Cowin [16]). However, the majority of bone adaptation models assume bone as an isotropic material with an equivalent elastic modulus given by, $E^{eq} = C \mu^p$, where C is a constant, μ the apparent density (or relative density) ant p an integer value. The works of Carter *et al.* [17] and Weinans *et al.* [4] are examples where this assumption is used. In such models the orientation of bone is not a variable due to the isotropy assumption. This limitation is overcome, for instance, by Cowin [16] through the introduction of a fabric tensor reflecting the bone anisotropy and by Jacobs *et al.* [9], where the "optimal" anisotropic material properties tensor is obtained for each local state of strain.

   In this chapter it is proposed an alternative methodology to account for both changes in relative density and orientation. It assumes trabecular bone as a cellular material obtained by the periodic repetition of a material cell with prismatic holes, this cell and its periodic repetition characterizes the so-called material microstructure (see Fig. 1). The relative density at each point depends on local hole dimensions, a_1, a_2 and a_3, *i.e.*, $\mu = 1 - a_1 a_2 a_3$. In addition, since the cellular material selected is strictly orthotropic, the model allows for the consideration of a variable orientation of unit cells as shown in Pedersen [18] and, consequently the simulation of bone as an oriented material. The elastic properties for this material are computed using the homogenisation method discussed below. Thus, at each point in the body, there is a microstructure characterised by the parameters $a = \{a_1, a_2, a_3\}^T$, defining the local material relative density, and by an orientation characterised by the Euler angles $\theta = \{\theta_1, \theta_2, \theta_3\}^T$.

   The assumption of a periodic microstructure as described above constrains the material symmetry to strictly orthotropic apparent material properties. It is important to note that this microstructure excludes isotropic symmetry in the apparent material properties. The highest material symmetry level it can represent is cubic symmetry, when $a_1 = a_2 = a_3$. It should also be emphasized that the periodicity is local and thus the density and orientation can vary along the domain.

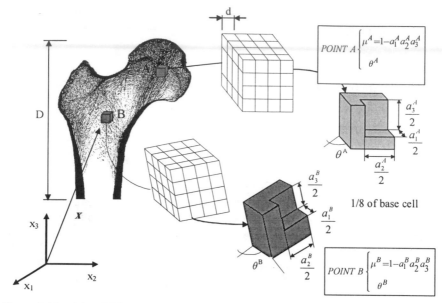

Figure 1. Material model for trabecular bone.

2.2   Homogenized elastic properties for trabecular bone

For the material model described, the homogenization method is a natural way to compute the bone effective (equivalent) properties. The method relies on the existence of two geometrical scales in the problem. The macro scale, that can be associated with a characteristic dimension of the bone (D), and a micro scale that characterises the dimension of the trabeculae (d) (see Figs. 1 and 2). Based on the scale parameter $\varepsilon = d/D$ it is possible to write the displacement field dependence on position, on a "macro" variable x and on a "micro" variable y defined as $y = x/\varepsilon$, reflecting the dependence on the microstructure.

The equilibrium for a deformable solid Ω with boundary Γ, made of the periodic material, subjected to body and surface loads b and t, respectively, is given by,

$$\int_{\Omega} F_{ijkl} \frac{\partial u_k^{\varepsilon}}{\partial x_l} \frac{\partial v_i}{\partial x_j} \Phi(y) d\Omega = \int_{\Omega} b_i v_i \Phi(y) d\Omega + \int_{\Gamma_t} t_i v_i d\Gamma \quad \forall v \; admissible \quad (1)$$

where E_{ijkl} is the tensor of elastic constants for base material (compact bone in the present model) and $\Phi(y)$ is a characteristic function defined by,

$$\Phi(y) = \begin{cases} 1 & se \quad y \in \yen \\ 0 & se \quad y \notin \yen \end{cases} \quad (2)$$

where ¥ is the solid part of base cell (figure 2), and u^f is the displacement field dependent on x e y.

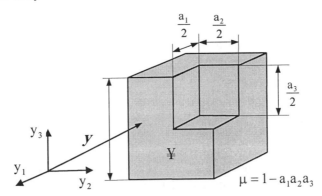

Figure 2. 1/8 of base cell.

Assuming an asymptotic expansion in the scale parameter ε,

$$u^\varepsilon\left(x,y\right)=u^0\left(x\right)+\varepsilon\,u^1\left(x,y\right)+\varepsilon^2 u^2\left(x,y\right)+\cdots,\ y=\frac{x}{\varepsilon} \qquad (3)$$

introducing it into the equilibrium Eq. (1) and taking the limit $\varepsilon \to 0$, one finds that the homogenised displacement field $u^0\left(x\right)$ is the solution of the equilibrium equation with the cellular locally periodic material replaced with an equivalent homogenised material.

In the case of a homogeneous base material (in our case the base material is compact bone), the homogenized (apparent) material properties are given by,

$$E_{ijkl}^H\left(\mu\right)=E_{ijkl}\,\mu-\frac{1}{|Y|}\int_Y E_{ijpm}\frac{\partial\chi_p^{kl}}{\partial y_m}dY \qquad (4)$$

as a function of the relative density $\mu=1-a_1a_2a_3$, where ¥ is the unit cell sub-domain occupied by homogeneous material (compact bone) as shown in figure 2. In the previous expression, the set of periodic functions χ^{kl} is solution of the set of equilibrium equations,

$$\int_{¥} E_{ijpm}\frac{\partial\chi_p^{kl}}{\partial y_m}\frac{\partial v_i\left(y\right)}{\partial y_j}dY=\int_{¥}E_{ijkl}\frac{\partial v_i\left(y\right)}{\partial y_j}dY,\ \ \forall v\ Y\text{-Periodic} \qquad (5)$$

defined on ¥.

The definition of the apparent mechanical properties (1) involves contribution of two terms. The first is only a function (actually a linear function) of the relative density and so is independent of the specific geometry of the material cell. The second (a correcting factor) introduces the influence of the material cell geometry via the six displacement functions, χ^{kl}, which are the solutions of a set of equilibrium equations (5) on the cell domain.

Besides the dependence on $\mu(a)$, the material properties depend also on the material orientation θ of the cells. This effect is taken into consideration by the rotation of homogenised material properties tensor, i.e.,

$$\left(E_{ijkl}^{H}\right)_{\theta} = R_{im}R_{jn}R_{kp}R_{lq}E_{mnpq}^{H} \tag{6}$$

where R is the coordinate transformation tensor associated to the cell rotation θ.

The homogenisation method applied to a natural cellular material, such as bone, is limited by the assumption of a local periodic microstructure. Since trabecular bone is not truly periodic, this assumption can only be understood as an approximation. However we should highlight the local character (in the y variable) of the periodicity assumption that leads, in the limit as $\varepsilon \rightarrow 0$, to a global model (in the x variable) without periodicity constraints. Furthermore the periodicity assumption yields two mechanically compatible subproblems (global and local problems) that can be solved separately thus making the bone model computationally treatable. The interested reader is referred to Sanchez-Palencia [19] for a full mathematical description of the homogenisation method. Equivalent mechanical models based on periodic assumptions have also been proposed for trabecular bone (see e.g. Gibson and Ashby [20]).

3   Bone Remodelling Model

3.1   Global optimization problem

To characterize the adaptation process let us assume that the bone adapts to the mechanical loading so that it tries to maximize its global stiffness. Obviously such a hypothesis just considers mechanical effects and leaves out numerous non-mechanical factors that are critical for the process. To correct this it is introduced a factor that will take these effects into consideration. So, unlike structural optimization models where the total amount of material is a priori prescribed [21], here such a constraint will be meaningless so he will assume that bone behaves like an open system where the total mass is regulated by a term that introduces into the objective functional the metabolic cost for bone formation.

Using a multiple load criterion, considering only applied surface forces and defining the vector of design variables $a = \{a_1, a_2, a_3\}^T$, the material optimisation problem can then be stated as,

$$\min_{a,\theta} \left\{ \sum_{P=1}^{NC} \alpha^P \left(\int_\Omega b_i^P u_i^P \, d\Omega + \int_{\Gamma_t} t_i^P u_i^P \, d\Gamma \right) + \kappa \int_\Omega (1 - a_1 a_2 a_3) \, d\Omega \right\} \tag{7}$$

subjected to,

$$0 \le a_i \le 1 \tag{8}$$

$$\int_\Omega E_{ijkl}^H (a,\theta) e_{ij} \left(u^P \right) e_{kl} (v) \, d\Omega - \int_\Omega b_i^P v_i \, d\Omega - \int_{\Gamma_t} t_i^P v_i \, d\Gamma = 0 \;\; \forall v = 0 \; in \, \Gamma_u \tag{9}$$

where E_{ijkl}^H are the homogenised material properties, e_{ij} is the strain field and v^P the virtual displacement for the pth load case. In the problem stated above, the first term of cost function is a weighted average of the structural compliance for each load case, where α^P are the load weight factors satisfying $\sum_{P=1}^{NC} \alpha^P = 1$, and NC is the number of applied load cases. The cost parameter κ in the second term plays an important role, since the resulting optimal bone mass will depend strongly on its value. It is known that even in the presence of identical loading conditions, the remodelling response is different for different individuals (Cowin [16]). Therefore, the parameter κ would include biological factors such as age, hormonal status and disease.

3.2 Necessary conditions for optimum – Law of bone remodelling

To obtain the necessary conditions characterizing the optimum lets consider the augmented functional (Lagrangian)

$$L = \sum_{P=1}^{NC} \alpha^P \left(\int_\Omega b_i^P u_i^P \, d\Omega + \int_{\Gamma_t} t_i^P u_i^P \, d\Gamma \right) + \kappa \int_\Omega \mu(a) \, d\Omega$$

$$- \sum_{P=1}^{NC} \left[\int_\Omega E_{ijkl}^H e_{kl} \left(u^P \right) e_{ij} \left(v^P \right) d\Omega - \int_\Omega b_i^P v_i^P \, d\Omega - \int_{\Gamma_t} t_i^P v_i^P \, d\Gamma \right] \tag{10}$$

$$- \int_\Omega \eta_{1i} (1 - a_i) \, d\Omega - \int_\Omega \eta_{2i} a_i \, d\Omega$$

where v^P, η_1 and η_2 are the Lagrange multipliers, associated with the equilibrium equation and lateral constraints, satisfying

$$v^P = 0, \qquad on \quad \Gamma_u$$

$$\eta_{1i}(x) \ge 0, \qquad \forall x \in \Omega \tag{11}$$

$$\eta_{2i}(x) \ge 0, \qquad \forall x \in \Omega$$

The first variation of the functional with respect to (w.r.t.) design variables, state variables and Lagrange multipliers, gives the necessary conditions for the problem.

The stationarity conditions w.r.t. the design variables a and θ give, respectively,

$$\sum_{P=1}^{NC}\left\{\int_{\Omega}\left[-\alpha^{P}\frac{\partial E_{ijkl}^{H}}{\partial a_{m}}e_{kl}\left(u^{P}\right)e_{ij}\left(u^{P}\right)\right]\delta a_{m}d\Omega\right\}+$$

$$+\int_{\Omega}\left(\eta_{1m}-\eta_{2m}\right)\delta a_{m}d\Omega+\kappa\int_{\Omega}\left(\frac{\partial\mu}{\partial a_{m}}\right)\delta a_{m}\ d\Omega=0\qquad\forall\delta a_{m}$$

(12)

and

$$\sum_{P=1}^{NC}\left\{\int_{\Omega}\left[-\alpha^{P}\frac{\partial E_{ijkl}^{H}}{\partial\theta}e_{kl}\left(u^{P}\right)e_{ij}\left(u^{P}\right)\right]\delta\theta\ d\Omega\right\}=0\ \forall\delta\theta$$

(13)

In the previous equations the adjoint field (Lagrange multipliers of the equilibrium constraints) v^{P} is substituted by the relation $v^{P}=-\alpha^{P}u^{P}$, that results from the comparison of the stationarity conditions with respect to u^{P} and v^{P}.

The cost function (7) is a global criterion, and the solution of the problem gives the stiffest trabecular bone structure, as described by its relative density and orientation (see figure 1) distribution for the given loads. The total bone mass is regulated, not only by the load values, but also by the parameter κ. The necessary conditions (12, 13) characterize the law of bone remodelling in the sense that whenever they are satisfied no remodelling will occur (remodelling equilibrium) and their solution leads to an optimal distribution of density and orientation for the stiffest trabecular bone.

3.3   Comparison with the local stimulus criterion models

Although it is not readily apparent, the global optimal criterion stated in eq. (12) is equivalent to a local stimulus criterion, such as the strain energy criteria employed by Beaupré et al. [22] and Weinans et al. [4]. In the case where the design variable is the relative density μ (not explicitly dependent on void dimensions a) the optimal condition can be rewritten in a local form as,

$$\frac{\partial E_{ijkl}^{ef}}{\partial\mu}e_{ij}\left(u\right)e_{kl}\left(u\right)-\kappa=0$$

(14)

for a single load case. The "ef" superscript indicates effective properties, not necessarily obtained by asymptotic homogenisation models. Assuming now that the trabecular bone is an isotropic material whose effective mechanical properties have a power law dependence on μ i.e.,

$$E^{ef}_{ijkl} = E^0_{ijkl} \mu^n \tag{15}$$

where E^0_{ijkl} are the base material (compact bone) constants and n an integer exponent, the derivative of the material properties with respect to relative density is,

$$\frac{\partial E^{ef}_{ijkl}}{\partial \mu} = n\, E^0_{ijkl} \mu^{n-1} = n\, \frac{E^{ef}_{ijkl}}{\mu} \tag{16}$$

Introducing the derivative (16) in the remodelling equation (14) one obtains,

$$\frac{U}{\mu} = \frac{\kappa}{n} \tag{17}$$

where U is the strain energy density, given by,

$$U = E^{ef}_{ijkl} e_{ij}(\boldsymbol{u}) e_{kl}(\boldsymbol{u}) \tag{18}$$

The remodelling rule presented in Weinans *et al.* [4] is,

$$\frac{\partial \mu}{\partial t} = B\left(\frac{U}{\mu} - K\right) \tag{19}$$

and so, at remodelling equilibrium $\left(\dfrac{\partial \mu}{\partial t} = 0\right)$ we have,

$$\frac{U}{\mu} = K \tag{20}$$

which is equivalent to equation (17) if one identifies $K = \kappa/n$.

This result means that assuming a polynomial approach for effective bone properties, instead of the homogenization method, the model is equivalent to the one presented by Weinans *et al.* [4]. However, considering a material with orthotropic microstructure instead of an isotropic material has the advantage that the material orientation is not imposed but results from the problem solution and would simulate the trabecular bone orientation.

3.4   Numerical model

The solution for the optimization problem is obtained solving the optimal conditions (12) and (13). The finite element method is used to compute the displacement field u^p, solution of equilibrium equation. The discretization to solve the equilibrium problem is also used to obtain the optimal distribution of bone density. With this assumption the cell dimension is considered constant within each finite element, thus the relative density is uniform within the elements.

To sum up, the necessary conditions are computationally solved in the following steps: First the homogenised elastic properties are computed for an initial solution. Next one computes the displacement field u^h due to the applied loads (the superscript 'h' denotes the solution obtained by the finite element method). Based on the displacement field finite element approximation, the necessary optimality conditions are checked. If they are satisfied the process stops, if not, improved values of the design variables are computed and the process restarts (see Fig. 3).

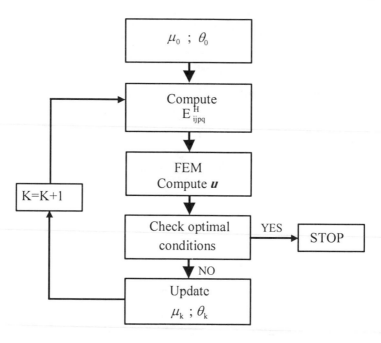

Figure 3. Flow chart for the computational procedure. K stands for iteration number.

The homogenised properties $E_{ijkl}^H (a)$ are obtained, for each optimisation iteration, by a polynomial interpolation on the interval $[0,1]^3$. This is accomplished with the values at the interpolation points (in the interval $[0,1]^3$) computed using the homogenisation code PREMAT (Guedes and Kikuchi [23]).

For each load case, the approximated solution u^h, is the solution of the equilibrium equation,

$$\int_\Omega E^H_{ijkl} e_{ij}\left(u^h\right) e_{kl}\left(w^h\right) d\Omega - \int_\Omega b_i w^h_i d\Omega - \int_{\Gamma_t} t_i w^h_i d\Gamma = 0 \quad \forall w^h \, admissible, \quad (21)$$

solved by the finite element method, using ABAQUS [24].

It is assumed that the design variables a_i are constant within each finite element. This allows one to write the optimality condition independently for each element. The solution can then be obtained with an iterative procedure based on a first order Lagrangian method. (Luenberger [25]). The formulas to update the cell design variables for each element 'e' at the k_{th} iteration are,

$$\left(a^e_i\right)_{K+1} = \begin{cases} \max\left[(1-\zeta)\left(a^e_i\right)_K,0\right] & if \ \left(B^e_i\right)_k \le \max\left[(1-\zeta)\left(a^e_i\right)_K,0\right] \\ \left(B^e_i\right)_k & if \max\left[(1-\zeta)\left(a^e_i\right)_K,0\right] \le \left(B^e_i\right)_k \le \min\left[(1+\zeta)\left(a^e_i\right)_K,1\right] \\ \min\left[(1+\zeta)\left(a^e_i\right)_K,1\right] & if \min\left[(1+\zeta)\left(a^e_i\right)_K,1\right] \le \left(B^e_i\right)_k \end{cases} \quad (22)$$

with $\left(B^e_i\right)_k = \left(a^e_i\right)_k + s\left(D^e_i\right)_k$.

In the previous update scheme, the vector D_K, the descent direction at the kth iteration, is the negative of the Lagrangian gradient with respect to the design variables a. The parameter $\zeta > 0$ defines the active upper and lower bound constraints and the real number 's' is the step length selected by the user. The component of the direction vector D for the design variable a^e_i (hole dimension i for the eth element) is (see eq. 12),

$$D^e_i = \sum_{P=1}^{NC}\left\{\int_{\Omega^e}\left[\alpha^P \frac{\partial E^H_{ijmn}}{\partial a^e_i} e_{mn}\left(\left(u^h\right)^P\right) e_{ij}\left(\left(u^h\right)^P\right)\right] d\Omega\right\} - \int_{\Omega^e} \kappa \frac{\partial \mu}{\partial a^e_i} d\Omega \quad (23)$$

Once the new values a^e_i are found, the optimal orientation is computed taking advantage of the analytical solution of the optimality condition (13) as described below.

In the two-dimensional case, as discussed in Pederson [18], optimality is achieved when the orthotropic material axes are aligned with the directions of principal stresses and strains. The three-dimensional case is discussed by Rovati and

Taliercio [26] and Cowin [27]. In the general case of an orthotropic material, a complete solution is not available. Although, this coalignment between principal strain directions and material orthotropy axes is a necessary condition, it does not guarantee sufficiency. Based on this fact, the computational model developed here assumes that the material directions of orthotropy are aligned with the principal strain directions for each individual finite element only if the new orientation corresponds to a better solution (globally stiffer). If this is not the case, the previous orientations are maintained.

A single load case was implicitly assumed in the previous conclusion that the optimal solution of the optimization problem must meet the condition of coalignment of the principal strains and the material directions. Indeed, the arguments that lead to the trajectorial theory of Wolff [1] seem to be based on consideration of a single load case. Determination of an optimal orientation of an orthogonal microstructure exposed to multiple loads is more problematic. In general, the principal strain directions associated with the various applied loads are not coaligned and, as such, an optimal solution based on coalignment is automatically ruled out for strictly orthotropic apparent material properties. Furthermore, as observed by Pidaparti and Turner [28] nonorthogonal microstructures are mechanically better suited to support multidirectional loading.

4   Numerical Examples for Bone Remodelling

To show the applicability of the model we will show two results from Fernandes *et al.* [13]. First the distribution of bone density and orientation were obtained for an intact femur. With this analysis it is possible to conclude if the model reproduces the behaviour of bone, and to estimate the value of k. Then, the model was applied to a femur implanted with a hip stem, in order to observe the bone remodelling when loads are redistributed due to the insertion of an implant.

4.1   Intact femur

A three-dimensional finite element model of the proximal femur was built using 5616 eight-node solid elements (see Fig. 4). The femur geometry is based on *Standardized Femur* (Viceconti *et al.* [29]). The applied forces are shown in table I and correspond to load cases used by Kuiper [30]. The first and the second case correspond to walking and the third corresponds to stair climbing. The problem was first solved for the single load case situation applying each load case individually, and then considering the multiple load criterion, for equal load weights, i.e., $\alpha_1 = \alpha_2 = \alpha_3$. An optimal solution was obtained starting with an initially homogeneous solution with $a_1 = a_2 = a_3 = 0.88$. Successive improvements to the solution were sought along the optimal search direction (23) followed by a realignment of each element's microstructure orientation, θ, with the directions of principal strain. It was assumed that the bone tissue material has the mechanical properties of compact bone. This means that dense compact bone corresponds to a cellular material with relative density equal to 1 and trabecular bone has values less than 1. We assumed a Young's modulus of 20 GPa for dense compact bone based on the experiments of Currey [31].

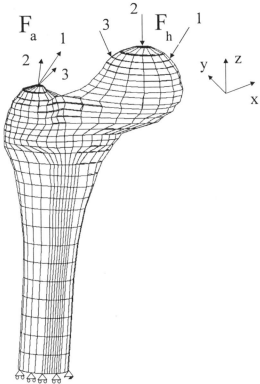

Figure 4. Three-dimensional model of an intact femur.

Table I - Load Cases				
LOAD		F_x (N)	F_y (N)	F_z (N)
1	F_a	768	726	1210
	F_h	-224	-972	-2246
2	F_a	166	382	957
	F_h	136	-630	-1692
3	F_a	383	669	547
	F_h	457	-796	-1707

A parametric study to analyse the influence of the parameter κ on total bone mass was undertaken for the single load case 2. A similar study was conducted to analyse the influence of load magnitude on the total bone mass. Results of these studies are shown in figure 5. It can be observed the total bone mass growing with load and diminishing with *k*.

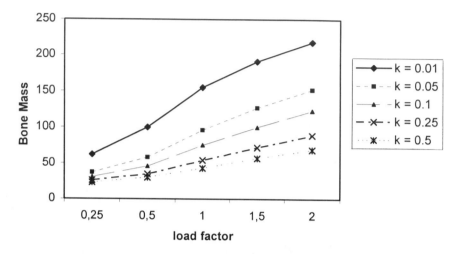

Figure 5. Parametric study of dependence of bone mass on *κ* and load factor.

The results obtained for multiple loads are presented in figures 6 and 7. After the optimal density distribution had been obtained, the orientation was computed for each load case independently applied to the final material distribution and superimposed in figure 8.

For each load case the obtained optimal orientations follow the principal strain directions and correspond well in regions of highly oriented trabecular bone in the saggital plane. In particular, there is evidence of the arching arcuate system of trabecular bone originating from the proximal ends of the lateral and medial cortices. Also they agree in general with those observed previously in two-dimensional models (*e.g.* Carter *et al.* [17] and Jacobs [9]).

Now lets analyse all the orientations concurrently. From the overlaid figure it is observed that in the region between the epiphyses and the diaphyses the orientations are roughly coincident. This means that the material has a preferential orientation. Therefore in this zone trabecular bone can not be isotropic while maintaining optimality. But, in the epiphyses the three orientations are significantly different. This is problematic, since it indicates that in this region the optimal trabecular bone microstructure should behave as an effective isotropic material to optimally support stresses from all directions and this cannot be achieved by a strictly orthotropic material.

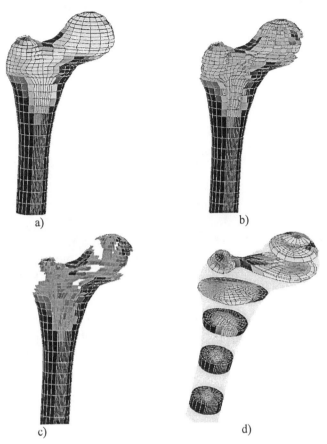

Figure 6. Bone density distribution for multiple load cases and κ=0.01. a) Whole femur. b) Elements with $\mu > 0.2$. c) Elements with $\mu > 0.4$. d) Femur cross sections.

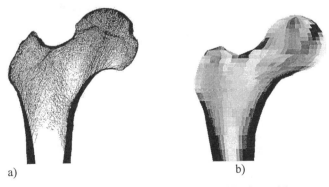

Figure 7. a) Real femur. b) Density distribution in femur obtained by the model.

Figure 8. Orientation for multiple loads.

4.2   Implanted femur

The global model for bone remodelling has also been applied to study the bone adaptation after a total hip arthroplasty. Figure 9 shows the finite element model for the implanted femur. The mesh was built using the femur geometry of the *Standardized Femur* (Viceconti *et al.* [29]) while the stem was modelled based on a Tri-Lock prosthesis from DePuy. The problem solved considers a total coated stem, resulting in a perfect adhesion of bone to metal. Mechanically the problem is modelled considering interface between bone and stem fully bonded. Thus there is continuity on the interface of two different materials. A more accurate model for bone/metal interface is possible, if one considers two bodies in contact. However the model should be extended to incorporate contact conditions. In Fernandes *et al.* [32, 33] a model for global bone remodelling with contact and bone ingrowth is described in detail.

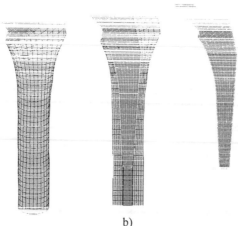

a)   b)   c)

Figure 9. Finite element model for an implanted femur. a) whole model b) saggital section c) stem geometry.

The problem was solved using the multiple load criteria, for load cases presented in table I with equal weights ($\alpha_1 = \alpha_2 = \alpha_3$). The remaining data are similar to the intact femur example. The initial solution is an uniform density distribution. The Young modulus for compact bone is 20 GPa, while for stem one considers 115 GPa, assuming a Titanium prosthesis. The value for metabolic cost is κ=0.01 N/mm². Figure 10 shows the bone density distribution. It is observed the bone loss on proximal femur due to stress shielding effect of the stem. This is clinical observed and also numerical obtained by different authors (see for instance Weinans *et al.* [34]).

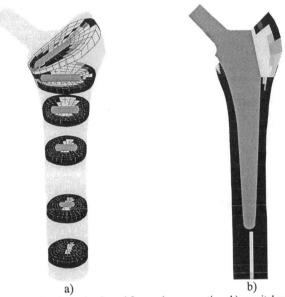

a)   b)

Figure 10. Density distribution for an implanted femur a) cross sections b) saggital section,

5   Local-Global Bone Remodelling Model

The global model for bone remodelling has shown a good behaviour in the simulation of bone adaptation with respect to bone density. However there are some limitations when the problem is to find the trabecular bone orientation. In fact, the orthogonal microstructure, obtained by the periodic repetition of a unit cell with prismatic holes, constraints the material symmetry, while the trabecular bone may actually be isotropic, transversely isotropic, orthotropic, or fully anisotropic depending on the bone region.

It is possible to combine the global remodelling model with a local optimization procedure to find the microstructure accounting for anisotropic characteristic of bone (Rodrigues *et al.* [15]). This model is presented in the following sections.

5.1   Problem formulation

For this problem the objectives are two: To identify the relative density distribution, and simultaneously to identify the single scale topology of the unit cell characterising the trabecular bone (cellular material).

Consider a linear elastic two-dimensional body subjected to body forces **b**, boundary tractions **t** and made of a cellular material simulated by a microstructure obtained by the periodic repetition of small square cells (see figure 11). For the sake of simplicity the model is two-dimensional.

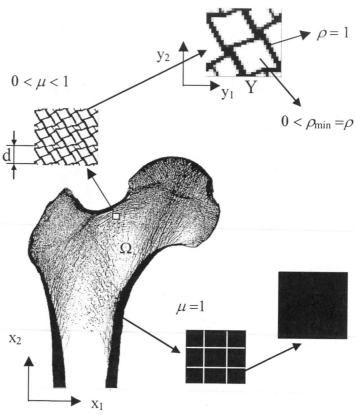

Figure 11. Global-local material model for bone remodelling.

The difference between this model and the previous material model is that in the present model the cell topology is unknown, thus the material symmetry is not imposed.

Similar to the global model, here one assumes the existence of two scales. A global (or macro) scale characterised by variable x and with a characteristic length equal to the structure (bone) global size (D) and a local (or micro) scale characterised by the variable y and with a characteristic dimension equal to the unit cell (trabecula) size d. At the micro (local) level, the cell is occupied with a material with variable density $\rho(y)$. Since the material distribution in the cell also depends on the global mechanical problem, actually the micro density ρ is a function of x and y.

For this model the relative density $\mu(x)$ at each point in the macro scale depends on the local distribution of the micro density ρ and is given by,

$$\mu(x) = \int_{Y(x)} \rho(x,y)\,dy \tag{24}$$

Note that $\mu(x)$ is the relative (apparent) density of trabecular bone at each point, $Y(x)$ the characteristic bone cell and $\rho(y)$ the density of material at trabecular (micro) level.

The optimization criterion for the problem is the same that was assumed for the global problem, i. e., bone adapts to form a structure with maximum stiffness for the applied loads.

Using a multiple load criterion the optimization problem can be stated as,

$$\min_{\substack{\mu \\ 0 < \rho(x,y) \le 1}} \left\{ \sum_{P=1}^{NC} \alpha^P \left(\int_\Omega b_i^P u_i^P \, d\Omega + \int_{\Gamma_t} t_i^P u_i^P \, d\Gamma \right) + \kappa \left(\int_\Omega \mu \, d\Omega \right) \right\} \tag{25}$$

subjected to,

$$\int_\Omega E_{ijkl}^H(\mu) e_{ij}(u^P) e_{kl}(v^P) \, d\Omega - \int_\Omega b_i^P v_i^P \, d\Omega - \int_{\Gamma_t} t_i^P v_i^P \, d\Gamma = 0, \ \forall v^P \ adm. \tag{26}$$

$$\mu(x) = \int_Y \rho(x,y)\,dY, \quad \forall x \in \Omega \tag{27}$$

where NC is the number of load cases and α^P the respective weight.

5.2 Homogenization

For this problem, the homogenisation method is also used to compute the equivalent material properties, since the assumption of periodic structure is still valid.

The base material in the unit cell depends on ρ, for this micro level it is assumed a polynomial dependence namely,

$$E = \rho^n E^0 \tag{28}$$

where E^0 is the elasticity tensor of compact bone and n an integer exponent. Note that this polynomial dependence is made at the unit cell level and not at the macro level. This implies that E^H is not constrained to have specific material symmetries.

Based on this assumption, the homogenised material properties, achieved in the limit $\varepsilon = d/D \to 0$, are defined by,

$$E_{ijkm}^{H} = \int_{Y} \left(\rho^{n} E_{ijkm}^{0} - \rho^{n} E_{ijpq}^{0} \frac{\partial \chi_{p}^{km}}{\partial y_{q}} \right) dy \qquad (29)$$

as a function of the material distribution, at the cell level $\rho(y)$. The periodic functions χ^{km} are solution of the set of local equilibrium equations,

$$\int_{Y} \rho^{n} E_{ijpq}^{0} \frac{\partial \chi_{p}^{km}}{\partial y_{q}} \frac{\partial w_{i}}{\partial y_{j}} dy = \int_{Y} \rho^{n} E_{ijkm}^{0} \frac{\partial w_{i}}{\partial y_{j}} dy, \quad \forall w \; Y - Periodic \qquad (30)$$

on Y, the material unit cell sub-domain (see Fig. 11).

Note that these relations are equivalent to ones given by Eqs. (4) and (5) with the base cell material given by function (28).

5.3   Numerical model

Defining the Lagrangian associated with the optimisation problem defined by Eqs. (25), (26) and (27), the stationarity condition w.r.t. the design variable $\rho(y)$ is,

$$\sum_{P=1}^{NC} \left[\alpha^{P} \frac{\partial E_{ijkl}^{H}}{\partial \rho(y)} e_{kl} \left(u^{P}(\hat{x}) \right) e_{ij} \left(u^{P}(\hat{x}) \right) \right] - \Lambda(\hat{x}) = 0 \qquad (31)$$

to be satisfied, for each fixed $\hat{x}$ in Ω, at all y's in the unit cell domain (see [15] for a complete derivation of this condition).

In the previous equation, $e\left(u^{P}(\hat{x}) \right)$ identifies the strain tensor, evaluated at $\hat{x}$, compatible with the global displacement field u^{P} solution of the equilibrium equation (26) and $\Lambda(\hat{x})$ is the Lagrange multiplier for the local relative density constraint (27).

From stationarity with respect to the design variable $\mu(x)$ one obtains the condition,

$$\Lambda(x) = \kappa \qquad (32)$$

In the equation (31) the adjoint fields (Lagrange multipliers of the equilibrium constraints (26)) v^{P} satisfy $v^{P} = \alpha^{P} u^{P}$. This derives from the stationarity conditions with respect to the state and adjoint displacement fields at equilibrium, u^{P} and v^{P}, respectively. Moreover, in the developments here the specific details of the homogenised mechanical properties derivatives w.r.t. design ρ have been omitted for clarity. The interested reader is referred to the work by Sigmund [35] where a detailed development is presented.

For the numerical implementation, global equilibrium equation (26) and local equilibrium equation (30) are solved via appropriate finite element approximations. The design variables global μ and local ρ are interpolated as constants in the elements of the respective meshes (global and local meshes respectively). This implies that the finite element mesh for the computation of the material microstructure is constant in each finite element of the global mesh, thus reducing substantially the computational cost.

Assuming this discretization for the design variables and using relation (32), the necessary condition (31) is approximated as,

$$\sum_{P=1}^{NC}\left[\alpha^{P}\,\frac{\partial E_{ijkl}^{H}}{\partial\rho(y)}\,\left\langle e_{kl}\left(u^{P}\right)\right\rangle_{El}\,\left\langle e_{ij}\left(u^{P}\right)\right\rangle_{El}\right]=k \tag{33}$$

for all y's in the material unit cell. In the previous necessary condition, the index 'El' ranges over all the global mesh finite elements and $\langle\ \rangle_{El}$ identifies the average operator applied to the respective global element strain field.

Assuming the local design variable ρ^{e} constant at each local mesh finite element, the optimality condition (33) can be uncoupled for each local design variable. From the discrete interpretation of this condition and introducing the upper and lower bound constraint thickness parameter ζ (defined by the user), the local solution is obtained by the fixed point method,

$$\left(\rho^{e}\right)_{i+1}=\begin{cases}\max\left[\left(1-\zeta\right)\left(\rho^{e}\right)_{i},\rho_{\min}\right] & \text{if } \gamma^{e}\left(\rho^{e}\right)_{i}\leq\max\left[\left(1-\zeta\right)\left(\rho^{e}\right)_{i},\rho_{\min}\right] \\ \min\left[\left(1+\zeta\right)\left(\rho^{e}\right)_{i},1\right] & \text{if } \min\left[\left(1+\zeta\right)\left(\rho^{e}\right)_{i},1\right]\leq\gamma^{e}\left(\rho^{e}\right)_{i} \\ \gamma^{e}\left(\rho^{e}\right)_{i} & \text{otherwise}\end{cases} \tag{34}$$

with the multiplier γ^{e} given by,

$$\gamma^{e}=\frac{\sum_{P=1}^{NC}\left[\alpha^{P}\,\dfrac{\partial E_{ijkl}^{H}}{\partial\rho(y)}\,\left\langle e_{kl}\left(u^{P}\right)\right\rangle_{El}\,\left\langle e_{ij}\left(u^{P}\right)\right\rangle_{El}\right]}{k} \tag{35}$$

In the previous algorithm, index 'e' ranges over all the finite elements in the local mesh and 'i' is the iteration counter.

Based on this iterative scheme the solution is obtained through successive steps between the global problem, to obtain the averaged global strain fields, and the local variable updating algorithm given by Eqs. (34) and (35). Note that for each new set of global strains only a limited number of iterations are taken in the local variable updating algorithm. A detailed description of this model and its numerical implementation is presented in Rodrigues *et al.* [15, 36].

5.4   *Two-dimensional results*

In this section results for the global-local bone remodelling are presented. These results were obtained for a two-dimensional finite element model of the proximal femur.

Three load cases are considered to simulate the loading history. The global finite element mesh has 1144 eight-node isoparametric elements and a side plate with 567 elements (see Jacobs *et al.* [9] for details on load conditions and global finite element model). The local mesh has 900 four node isoparametric elements. In the computational model it was considered the exponent 'n' in Eq. (28) equal to 4. A high value for 'n' leads to cell topologies with few intermediate densities, which is convenient for a correct identification of trabecular bone geometry. Figure 12 shows the relative density distribution and the final microstructure for selected elements.

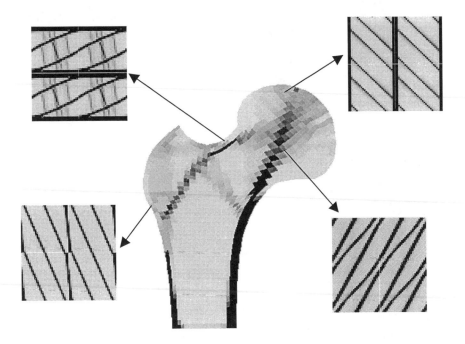

Figure 12. Results for global-local bone remodelling model. Optimal density distribution and local microstructures for selected elements.

6 Discussion

In this chapter two models based on optimization strategies to simulate computationally the bone remodelling process were described. A global model where bone relative density and orientation are obtained by an optimization problem but with fixed material symmetry and a global-local model which introduces the material symmetry as a design variable.

The bone is assumed as a cellular material with period microstructure. For the global model the microstructure is obtained by the periodic repetition of a unit cell with prismatic holes. Such approach constraints the material symmetry to a strictly orthotropic material. For the global-local model, a periodic microstructure is still assumed, however the base cell topology is not imposed *a priori* but results from the remodelling process.

For both models, the optimization objective function is the minimization of a linear combination of the compliance (inverse of stiffness) and the metabolic cost of bone apposition. Thus, the problem combines the purpose of increasing mechanical performance without increasing the biological cost associated to bone maintenance. The metabolic cost of bone maintenance is given by the κ parameter, which is constant in this model but may be generalised to a non-linear function of density and location and that could be used to characterise biological factors such as disease, hormonal status, or age. As demonstrated in the parametric study performed the total amount of bone mass depends strongly on this parameter .

The global model was applied to a three-dimensional model of a femur, discretised by eight-node solid elements, with the objective of comparing the results with the density distribution and trabecular orientation of real bone. The results obtained for the density distribution reflected several morphological features observed in actual density patterns.

An important issue of the model is the adoption of a microstructure for trabecular bone. This feature permits to simulate both changes in relative density and orientation. However, for multiple load cases this is a over constrained model since it is difficult to obtain an optimally oriented orthogonal microstructure. This problem is overcome with the global-local model where the optimal cell topology is obtained.

With respect to relative density the consideration of multiple load cases is critical to the prediction of certain morphological features. For instance the thickness of distal cortex is not uniform. The predicted structure was stronger in bending about the y axis than about the axis x and in torsion. This result emphasizes the importance of a multiple load formulation. Also the influence of the out of plane loads can only be captured with a three-dimensional model such as the one presented in this chapter.

From the orientations obtained, it is apparent that an orthotropic microstructure for trabecular bone of the proximal femur can only be optimal for a single load case. This is a suggestion from the observation that in some regions the optimal orthotropic orientation was found to be different for different load cases. Here we note that this was observed with only three loads. It probable that this non homogeneity in optimal material symmetry will be more evident as one goes to real applications were one should consider more loadings. From the optimal orientations obtained here one can conclude that the optimal material symmetry of trabecular bone varies from region to region, and may be isotropic, transversely isotropic, orthotropic, or anisotropic depending on the orientations of the local mechanical loadings it must withstand. To overcome this, the global-local model obtains the optimal microstructure characterizing the equivalent global properties without imposing any material symmetry. It should be noted that the objective of obtaining a microstructure is not to reproduce geometrically the trabecular architecture, but to model and to approximate its equivalent properties, namely material density, orientation and symmetry.

The global model was also applied to an implanted femur. In this case it was considered a fully bonded interface between the stem and the bone, i.e. a totally coated stem where perfect bone ingrowth occurred in the stem surface. Results show bone loss on the proximal region, due to stress shielding effects. These results are clinically observed and agree with results of numerical studies presented by other authors. However a more precise model for bone interface can be considered, namely modelling the bone and stem as two bodies in contact. In that case, the model has to be adjusted to incorporate contact effects (see e.g. Fernandes *et al.* [32, 33]).

The optimization models can give an important contribution to the understanding of the biomechanics processes in human body, namely the bone behaviour as an adaptive structure. Furthermore, these models permit not only the development of consistent mechanical and mathematical models for bone remodelling but also they can have an important contribution in the identification of the respective stimulus.

Acknowledgements

The authors want to acknowledge and thank Prof. José Miranda Guedes, and Eng. João Folgado, IST - Technical University of Lisbon, Portugal, Prof. Cristopher Jacobs, Stanford University, USA, and Prof. Martin Bendsøe, Technical University of Denmark, for a very rewarding collaboration in the development of the models and results presented in this chapter. The work described here was supported by the Portuguese Foundation for Science and Technology through projects POCTI/38367/EME/2001 and POCTI/35983/EME/1999.

References

1. Wolff J., *Das Gesetz der Transformation der knochen*, (Hirchwild, Berlin,1892). Translated as: *The Law of Bone Remodelling*, (Springer-Verlag, Berlin, 1986).
2. Treharne R. W., Review of Wolff's Law and its Proposed Means of Operation. *Orthopaedic Review* **10** (1981) pp.35-47.
3. Fyhrie D. and Carter D., A Unifying Principle Relating Stress to Trabecular Bone Morphology. *Journal of Orthopaedic Research* **4** (1986) pp. 304-317.
4. Weinans H., Huiskes R. and Grootenboer H. J., The Behavior of Adaptive Bone-Remodeling Simulation Models. *Journal of Biomechanics* **25** (1992) pp. 1425-1441.
5. Mullender M. G., Huiskes R. and Weinans H., A Physiological Approach to the Simulation of Bone Remodeling as a Self-Organisational Control Process. *Journal of Biomechanics* **27** (1994) pp. 1389-1394.
6. Brown, T. D. and Fergurson, A. B., Mechanical property distributions in the cancellous bone of the human proximal femur, *Acta Orthop. Scand.* **51** (1980) pp. 429-437.
7. Yang, J., Kabel, J., Rietbergen, B. Odgaard, A., Husikes, R. and Cowin, S. C., The Dependence of the Elastic Constants of Cancellous Bone Upon Volume Fraction, in *Synthesis in Bio Solid Mechanics*, P. Pedersen and M. P. Bendsøe Eds. (Kluwer, 1999).
8. Odgaard A., Kabel, J, Rietbergen, B. and Huiskes R., Relations Between Architectural 3-D Parameters and Anisotropic Elastic Properties of Cancellous Bone, in *Synthesis in Bio Solid Mechanics*, P. Pedersen and M. P. Bendsøe Eds. (Kluwer, 1999).
9. Jacobs C., Simo C., Beaupré G. and Carter D., Adaptive Bone Remodelling Incorporating Simultaneous Density and Anisotropy Considerations. *Journal of Biomechanics* **30**(6) (1997) pp. 603-613.
10. Hollister S., Kikuchi N. and Goldstein S., Do Bone Ingrowth Processes Produce a Globally Optimized Structure? *Journal of Biomechanics* **26**(4/5) (1993) pp. 391-407.
11. Garcia J. M. Martinez M. A. and Doblaré M., An Anisotropic Internal-External Bone Adaptation Model Based on a Combination of CAOS and Continuum Damage Mechanics. *Computer Methods in Biomechanics and Biomedical Eng..* **4**(4) (2001) pp. 355-378.
12. Doblaré M. and Garcia J. M., Application of a Bone Remodelling Model Based on a Damage-Repair Theory to the Analysis of the Proximal Femur Before and After Total Hip Replacement. *Journal of Biomechanics* **34**(9) (2001) pp. 1157-1170.
13. Fernandes P., Rodrigues H. and Jacobs C., A model of bone adaptation using a global optimization criterion based on the trajectorial theory of Wolff.

Computer Methods in Biomechanics and Biomedical Engineering, **2** (1999) pp. 125-138.

14. Pidaparti, R. M. V. e Turner, C. H., Cancellous bone architecture: Advantages of nonorthogonal trabecular alignment under multidirectional joint loading. *J. Biomechanics* **10** (1997) pp. 979-983.

15. Rodrigues H., Jacobs C., Guedes, J. M., Bendsøe, M.P., Global and Local Material Optimization Models Applied to Anisotropic Bone Adaptation, in *Synthesis in Bio Solid Mechanics*, P. Pedersen e M. P. Bendsøe Eds. (Kluwer, 1999) pp. 209-220.

16. Cowin S.C., Bone Mechanics, (CRC Press, Boca Raton, Fl., 1989).

17. Carter D., Orr T. and Fyhrie D., Relationships Between Loading History and Femoral Cancellous Bone Architecture, *Journal of Biomechanics* **22**(3) (1989) pp. 231-244.

18. Pedersen P., On Optimal Orientation of Orthotropic Material, *Structural Optimization* **1** (1989) pp. 101-106.

19. Sanchez-Palencia E., Non-Homogeneous Media and Vibration Theory, *Lecture Notes in Physics 127* (Springer, Berlin, 1980.)

20. Gibson L. J. and Ashby M. F., Cellular Solids- Structures & Properties, (Pergamon Press, 1988).

21. Bendsøe M. P. and Kikuchi N., Generating Optimal Topologies in Structural Design using a Homogenization Method, *Computer Methods in Applied Mechanics and Engineering* **71** (1988) pp. 197-224.

22. Beaupré G. S., Orr T. E. and Carter D. R., An Approach for Time-Dependent Bone Modeling and Remodeling - Theoretical Development, *Journal of Orthopaedic Research* **8** (1990) pp. 651-661.

23. Guedes, J. M. and Kikuchi, N., Preprocessing and Postprocessing for Materials based on the Homogenisation Method with Adaptive Finite Elements Methods, *Computer Methods in Applied Mechanics and Engineering* **83** (1990) pp. 143-198.

24. ABAQUS, User's Manual, Version 5.8, (Hibbit, Karlsson & Sorensen, Inc, RI, USA.1998).

25. Luenberger D. G., Linear and Nonlinear Programming, (Addison-Wesley, 2ª edition , 1989).

26. Rovati M. and Taliercio A., Optimal Orientation of the Symmetry Axes of Orthotropic 3-D Materials, in *Proc. Int. Conf. Engng. Optimization in Design Processes, Karlsruhe; Lecture Notes in Engineering, 63*, Eschenauer H. A., Mattheck C. e Olhoff N. Eds. (Springer Verlag, Berlin 1991) pp. 127-134.

27. Cowin, S., Optimisation of the Strain Energy Density in Linear Anisotropic Elasticity, *Journal of Elasticity* **34** (1994) pp. 45-68.

28. Pidaparti, R. M. V. e Turner, C. H., Cancellous bone architecture: Advantages of nonorthogonal trabecular alignment under multidirectional joint loading. *J. Biomechanics* **10** (1997) pp. 979-983.

29. Viceconti M., Casali M., Massari B., Cristofolini L., Bassini S. and Toni A., The 'Standardized Femur Program' Proposal for a Reference Geometry to be Used for The Creation of Finite Element Models of the Femur, *Journal of Biomechanics* **29**(9) (1996) pp. 1241.
30. Kuiper J. H., Numerical Optimization of Artificial Hip Joint Designs, Ph.D. Thesis (*Katholieke Universiteit Nijmegen*, The Netherlands, 1993).
31. Currey J., The Mechanical Adaptation of Bones (Princeton University Press, 1984).
32. Fernandes P., Folgado J., Jacobs C. and Pellegrini V., A Contact Model for Bone Remodelling on Total Hip Arthroplasty, *In Computer Methods in Biomechanics & Biomedical Eng. – 3,* Middleton J., Jones M. L., Shrive N. G. and Pande G. N. Eds. (2001) pp. 123-128.
33. Fernandes P. R., Folgado J., Jacobs C and Pellegrini V., A contact model with ingrowth control for bone remodelling around cementless stems, *Journal of Biomechanics* **35**(2) (2002) pp. 167-176.
34. Weinans H., Huiskes R. and Grootenboer H. J., Effects of Material Properties of Femoral Hip Components on Bone Remodeling, *J. Orthopaedic Research* **10**(6) (1992) pp. 845-853.
35. Sigmund, O., Materials with prescribed constitutive parameters: An inverse homogenization problem, *Int. J. Solids Struct.* **31** (1994) pp. 2313-2329.
36. H. Rodrigues, H., Guedes, J. M. and Bendsøe, M. P., Hierarchical optimization of material and structure, *Structural Optimization* **24** (1) (2002) pp. 1-10.

ANALYSIS OF THE BLOOD FLOW IN THE ARTERIES BY MEANS OF HAGEN - POISEULLE'S MODEL

JÓZEF WOJNAROWSKI

Department of Mechanics, Robots and Machines, Silesian Technical University, 18A Konarskiego St., 44-100 Gliwice, Poland

In this paper, the influence of such blood state parameters have been analysed as the number of hematocrit and plasma proteins, on the flow in the blood vessels. The arterial blood flow has been described in the form of Hagen-Poiseuille's model. The distributions of velocity, pressure drops and the hydrodynamic friction coefficient were determined by the application of the classic Wormesly equation and empirical non-Newtonian blood rheology model. The performed analysis shows the qualitative effects in changes of the rheological blood properties and their influence on the flow image.

1 Introduction

Investigations, which cover problems of the class of flows, nowadays named Hagen-Poiseuille's problems, have for many years been strongly connected with hemodynamical researches [1,2]. Almost 20-years ago Jean Luis carried out blood pressure measurements. M. Poiseuille using a mercury manometer established in a vessel new directions of studies in hydrodynamics. These investigations had an essential practical meaning. The result of the 30 years of Poiseuille's time-consuming tests was the empirical formula in which changes of pressure Δp, the flow rate (flow intensity, volume flux) $\dot{V}$, length l and diameter d of a channel were connected in the following way:

$$\dot{V} = K \frac{\Delta p d^4}{l} \qquad (1)$$

but he did not establish the physical meaning of the proportionality coefficient in this formula. He only pointed out a possible connection with temperature changes.

This formula was also obtained by Wiedeman and Hagenbach who independently published in 1856 and 1860 that:

$$K = \frac{\pi}{128\eta} \qquad (2)$$

i.e. as expected by Poiseuille, it is a parameter depending on temperature since it is a function of the dynamic viscosity coefficient η.

Therefore it is reasonable according to the formula, i.e. Hagen - Poiseuille's flow used nowadays, another name should be assigned - namely Gotthilf Hagen. Despite the fact that Poiseuille first considered the connection (1), Hagen also confirmed experimentally that:

$$\dot{V} \sim d^4 \qquad\qquad (3)$$

The Hagen - Poiseuille's rule is considered at present as so obvious that its importance cannot be overestimated. It should be admitted that the problem was, among others, considered by Claude Luis Navier, recognised as the founder of fluid mechanics, who established that:

$$\dot{V} \sim d^3 \qquad\qquad (4)$$

but as we know - this was wrong.

The model of Hagen - Poiseuille's flow provides a theoretical and basic, but very useful theory for the analysis of flow phenomena in blood vessels. This model is confirmed by the works of Wormersly [3,4], McDonald [1] or more recently by Rodkiewicz [5].

In the present paper the Hagen - Poiseuille's model has been used to analyse the influence of human blood state parameters, such as the hemocritic number and level of protein in the plasma (more precisely the fibrinogen and globulin level), on the behaviour of the velocity field in the cross-section of blood vessels. The values of pressure drop that arises during the blood flow through a vessel are of great importance. The formulation of the problem allows us to solve it analytically. Simultaneously, it allows focusing on the connection between the flow and hematological parameters which are the source of several uncertainties and controversies.

2   Circulation, Physiological Functions and Constitution of the Blood

Blood circulates in a vessel system, consisting of mutually connected channels with a tree structure. Two closed circulation cycles can be distinguished:

- the pulmonary circulation cycle, which forms a not very dense branching system of vessels, which permits to exchange gases with the external environment. It has its origin in the arterial cone of the right heart chamber (vertical) and ends in the left atrium (auricle);
- the great circulation cycle, which assures blood distribution throughout the human body. It has its origin in the single aortal artery starting at the arterial cone of the left   heart chamber and returns to the heart through the main veins entering the right heart auricle.

In both circulation cycles, blood flows from the heart chambers through arteries which branch off in arterioles and arterial capillaries. From the vast network of arterial capillaries, blood is collected via venous capillaries and veinlets through veins into the main veins which lead to the heart vestibules. Overview classifications of blood vessels due to their diameter and total area of the cross-section are presented in figure 1 [1]:

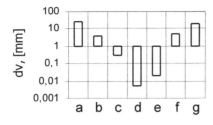 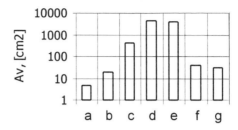

Figure 1. Overview classifications of blood vessels due to the diameter d_v of the vessel and the total area of the cross-section A : a) main artery (aorta), b) arteries, c) arterioles, d) arterial and vein capillaries, e) veinlets, f) veins, g) main veins.

Blood, is a fluid of particular smell, red colour and slight alkaline reaction (*pH = 7.35 ... 7.45*). It consists, in average, about 8% of the human body mass, i.e. 4...6 litres (in men and women, respectively).

From a physiological point of view, blood circulation has three essential functions:

- distribution (transport), which consists in the delivery and passing of particular chemical substances to and from every single cell of the body;
- control, which consists in maintaining stable physiological conditions inside the human body, in particular - temperature;
- defence, which consists in the detection and elimination of extraneous bodies and toxic substances and protection against exsanguinations.

By means of whirling, it is easy to distinguish the liquid phase of blood, called plasma, from the solid phase, i.e. blood cells or blood bodies (corpuscles) in form of red blood cells, white blood cells and thrombocytes. A general systematisation of the blood phases and types of their morphological components is presented in figure 2 [6].

Erythrocytes, also called red blood cells, are bi-concave cells whose physiological function consists in the transportation of breathing gases to and from the body cells. These are the most frequent blood components (in average there are *4,000,000 to 11,000,000* per mm^3) which makes 95% of all the blood-cell quantity [7]. Leukocytes, also called white blood cells, are: neutrophiles, eozynophiles, basophiles, lymphocytes, monocytes. The purpose of leukocytes is to protect the human body via phagocytosis and the production of antibodies. In contradiction to other morphocytological components they are full cells and possess the feature of active self-moving. Their average quantity in human body ranges from *4,000 to 10,000* per mm^3 of blood [7].

The last group of blood cells consists of thrombocytes (plate blood cells) whose function is protection against exsanguinations. In case of danger thrombocytes release the content of follicles causing a clamp of smooth muscles and initialization of blood clotting. The quantity of thrombocytes usually varies between *250,000 and 500,000 per mm* [1,7].

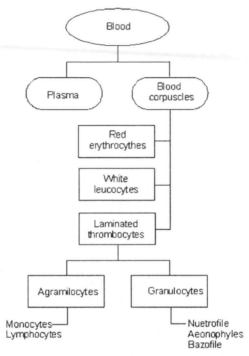

Figure 2. Scheme of blood phases and blood components.

In general, the diameter of human blood cells varies from 2μm to almost 20 μm. The smallest components are thrombocytes; the largest monocytes - which are considered as leucocytes. A comparison of the dimensions of the particular blood components is presented in figure 3 [8,6].

Plasma is a liquid phase of blood. It is a yellow-green fluid, which consists of 90% of water. Besides water, the main components of plasma are proteins (in average from *6.0 to 8.4 g/100ml*). By means of electrophoresis we distinguish albumins, globulins (α_1, α_2, β) and fibrinogen [6]. The contents of proteins are in average, the 7% of the plasma. The remaining 3% consists of various chemical organic and nonorganic substances, but they only influence the physical properties of blood only slightly.

The overall volume of the solid phase constitutes a very important parameter to describe the blood state from both the medical and fluid – mechanical (especially rheological properties) point of view. This parameter is called the hematocryte or hematocrytic number and is measured in volume per cent, i.e. *ml/100 ml*. Its average value is equal to *41 ml/100 ml* in the case of women, and *46 ml/100 ml* for men [6].

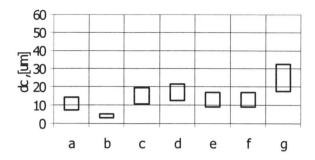

Figure 3. Comparison of the dimensions of human blood cells: a) erhytrocyte b)thrombocyte, c) neutrophile, d) eozynophile, e) basophile, f) lymphocyte, g) monocyte.

Morphology and chemical constitution are important blood parameters from the medical point of view and are considered to be essential data in diagnosis. Simultaneously, these parameters, (especially the hemocritic number, fibrogen content and globulins of plasma) are of essential importance for the rheological properties, influencing the form and evolution of the flow fields in vessels of the circulatory system.

3 Rheological Properties of Human Blood

For many fluids, the relation between deformation and stress states is linear; therefore, simple shearing tangent stresses τ can be expressed as a function of shear velocity $\dot{\gamma}$ by means of the Newtonian rule:

$$\tau = \eta \dot{\gamma} \tag{5}$$

where the proportionality coefficient η, called dynamic viscosity coefficient, is a material constant. Physiological liquids, blood in particular, the equation (5) is no longer valid:

$$\frac{\tau}{\dot{\gamma}} \neq const \tag{6}$$

Plasma, the liquid phase of blood, is from the rheological point of view a Newtonian fluid. At a temperature its viscosity coefficient of $37°C$, is 1.2 cP^2. Similarly, as in the case of every Newtonian fluid, changes of temperature involve changes of viscosity [9]:

$$\eta = \eta_0 e^{\frac{4200(T-T_0)}{(T+162)(T-162)}} \tag{7}$$

but physiologically these changes are insignificant. As has been shown, based on experimental investigations, changes of the viscosity coefficient affect changes in the weight concentration of proteins c_p, in particular of fibrinogen and globulines. The general form of this dependence is usually presented by means of the following formula [10]:

$$\frac{\eta}{\eta_w} = \frac{1}{1-k_p C_p}$$

(8)

where k_p is an experimentally determined constant and η_w denotes the viscosity coefficient of water. Nevertheless, blood is a polydispersional system because it contains suspended particles, i.e. blood cells. As can be seen from special experiments, adding even a slight quantity of (dispersed) solid phase into a fluid always causes an increase of its viscosity. Albert Einstein presented an analysis of these phenomena by means of theoretical investigations almost one hundred years ago. Considering a suspension of spherical particles with the volume density c_v in the continuous phase with the viscosity coefficient η_f, Einstein determined the viscosity coefficient of the suspension in the following form:

$$\frac{\eta}{\eta_f} = 1 + k_1 c_v$$

(9)

The coefficient k_1 depends on the shape of the suspended particles and their deformability. Einstein determined theoretically that in the case of stiff spheres $k_1 = 2.5$. Further investigations, it have found that restrictions of Einstein's formula are connected not only with the shape and deformability of the particles themselves, but also, first and foremost, with the concentrations of particles, in this case not greater than 1%. Attempts to extend the scope of the application of formula (9) have added terms of a higher order:

$$\frac{\eta}{\eta_f} = 1 + k_1 c_v + k_2 c_v^2 + \dots\dots$$

(10)

as for example in Thomas's formula [11]:

$$\frac{\eta}{\eta_f} = 1 + 2{,}5_1 c_V + 10{,}05_2 c_V^2 + 0{,}00273 e^{16{,}6 c_V}$$

(11)

its adequacy to the modelling of blood was restricted.

The true reasons of these defects were the pseudo-plastic blood properties as exposed by Dintenfass [12] on the turn of the 50's and 60's of the previous century. This concept was developed and experimentally verified several years later by Chien [13]. In the experiment, he investigated changes of reduced (relative) values of the viscosity coefficient for three blood-like specimens:

- **NP** - human blood with a proper constitution with the hematocrit number $H = 45$ *ml*/ 100 ml and viscosity coefficient of the plasma $\eta_f = 1.2$ *cP*;

- **NA** - in form of a suspension of erythrocytes in Ringer's 11% albumine solution; in this case the liquid phase was without globulines and fibrinogen (factor I) and the blood lost its ability to aggregate;
- **HA** - was obtained from the suspension NA by additional hardening of the erythrocytes. In result the blood not only lost its ability to aggregate, but also the suspended phase was characterised by stiff and non-deformitable particles.

Results of this experiment, in the form of charts of viscosity coefficients η_r are presented in figures 4, 5 and 6.

If blood were a mineral substance, the suspended phase of which consisted of stiff and non-deformable particles without the ability to create macroscopic clusters (as in the case of HA specimen), then its rheology could be described by means of more or less complicated formulas in the form of (9) or (10).

But the ability to aggregate erythrocytes, as has been stated by the co-operation of globulines and fibrinogene (compare courses NA and NP in figures 5 and 6) causes that erythrocytes in the range of low shear velocities are subjected to aggregation. Thus special 3-dimensional structures are created [14]. Together with an increase of the shear velocities, the gradual network detcriorates and the erythrocytes are reoriented along the main flow directions.

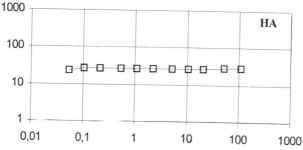

Figure 4. Distribution of the relative viscosity coefficient η_r as a function of shearing velocity γ for the suspension HA.

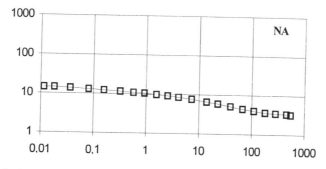

Figure 5. Distribution of the relative viscosity coefficient η_r as a function of the shearing velocity γ for the suspension NA.

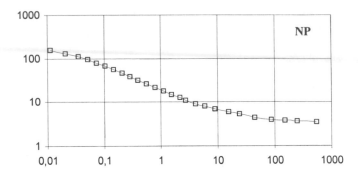

Figure 6. Distribution of the relative viscosity coefficient η_r as a function of the shearing velocity γ for the suspension NP.

As a consequence of the above-mentioned facts, the constitutive blood model should take into account the following three parameters:

- volume concentration of the solid phase c_v, which is given in the terms of the hematocritic number H (usually $c_v \approx 0.96H$, but some disturbances may occur depending on the assumed measurement methodology). This number determines the accuracy and frequency of collisions of adjoining blood cells;
- shear velocity γ, its value influences the dynamical creation and destruction of spatial aggregates of blood cells;
- physical-chemical properties of the liquid phase, but in this case the rheological properties are not so important. For blood plasma these are precisely determined. The other features connected with the aggregation of the suspended phase are more important.

Probably more importants are the contents of fibrinogen and globules. Additionally, the fourth group of parameters can be put on the list:

- properties of erythrocytes describing their ability to aggregate, in particular parameters connected with the adsorption and electrical potential of the membrane;
- but they should be recognised as a subject of future investigations. Due to the fact that the knowledge of the whole processes is not extensive enough to create an adequate model, it is not possible to formulate precise constitutive dependences of the blood.

A good example of a blood rheology model, taking into account two of the afore-mentioned parameters, was proposed by Scott Blair [15]:

$$\tau^{\frac{1}{2}} = \tau_0^{\frac{1}{2}} + \left(k_c \, \dot{\gamma} \right)^{\frac{1}{2}} \tag{12}$$

or

$$\tau^{\frac{1}{2}} = \tau_0^{\frac{1}{2}} + \eta_c \, \dot{\gamma}^{\frac{1}{2}} \tag{13}$$

previously used by Casson for oil printer's ink[16]. The coefficients k_c and η_c (called Casson's coefficients) used in formulas (12) and (13) are empirical constants of the model. They depend on the concentration of the scatter phase, i.e. hematocritic number in the case of blood. In figure 7, the course of the flow curve according to Casson's model is presented for the range of shear velocity $\dot{\gamma} = 0 \dots$ 20 $1/s$. The dependence $\tau(\dot{\gamma})$ for the Newtonian model with $\eta = 3.0$(Fig. 7(a)) and $\eta = 4.0\ cP^{20}$ (Fig. 7(b)) is also shown [2].

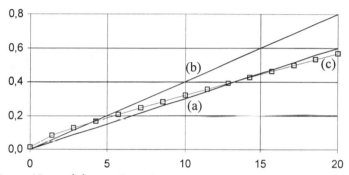

Figure 7. Curve of flow $\tau(\gamma)$ according to Casson's (c) and the Newtonian ($\eta = 3.0$ (a) and $\eta = 4.0$ cP (b)) models.

A slightly controversial parameter, used by Casson, is the flow limit τ_o, which is in the case of blood very difficult to determine experimentally. According to Scott Blair's proposal $\tau_o = 0.04$ dyn/cm^2, but it was obtained by extrapolation from the flow curve assuming Casson's model. Nowadays it is considered as a small value, and probably $\tau_o = 0.04$ dyn/cm^2 [14].

Experimental investigations show a very good adequacy of Casson's model for low shear velocities, but its a very bad approximation in the case of high γ values when the pseudo-plastic effects gradually disappear (compare Figs. 4,5 and 6). In consequence, Mettill and his co-workers presented in the late 60's a modification of the model in the same way as Merrill et al. [17]:

He restricted the usage of model equations (12) and (13) only to the case when $\dot{\gamma} \leq 20$; for $\dot{\gamma} = 20.0 \dots 100.0$ he proposed a linear transition:

$$\tau = A\dot{\gamma} + B \tag{14}$$

And for $\dot{\gamma} > 100$, the linear Newtonian model according to equation (3) was applied.

Such an elementary extended application of Casson's model showed to be so effective that Casson-Merrill's model has been applied until now (see work by R. Eilers, Chr. Ritter and G. Rau [18], where coefficients of Casson-Merrill's model considered here are listed).

An evident weakness of Casson-Merrill's model is the fact that only the shear velocity $\dot{\gamma}$ is apparently taken into account. The influence of the volume consistency of the scatter phase is represented though the coefficient k_c (or η_c), for the dependence $k_c = f(c_v)$ which is unknown (the values presented in references are connected only with the hematocrit number of 40ml/ 100ml). Simultaneously, in the model other parameters (determining the blood rheology) are completely neglected. For example, the level of fibrinogen and plasma globulines has not been taken into consideration.

Many examples and attempts to overcome the restrictions connected with Casson's equation or even totally new formulations of a blood-rheology model. A particularly vast, comprehensive review of the related problems was presented by Quemad in [14]. As a curiosity Casson's proposal might be mentioned. It was presented in the late 70's of the previous century (i.e. twenty years later than equation 10):

$$\frac{\eta}{\eta_f} = \frac{1-\alpha c_V}{(1-1,75\alpha c_V)^2} \tag{15}$$

Nevertheless, one of few formulations, in which the shear velocity γ, the hematocritic number H and fibrogene and $TPMA$ globuline level have been taken into account, is Walburn-Schneck's model [19]:

$$\tau = c_1 e^{c_2 H + c_4 \frac{TPMA}{H^2}} \cdot \gamma^{1-c_3 H} \tag{16}$$

The values of the empirical constants were determined by the authors as $c_1 = 0.797$, $c_2 = 0.0608$, $c_3 = 0.00499$, $c_4 = 145.85$. The shape of the flow curve according to Walburn-Schneck's model (equation 16) is presented in figure 8 along with tow curves for the Newtonian model whit $\eta= 3.0$ (Fig. 8(a)) and $\eta = 4.0$ cP (Fig. 8(b)).

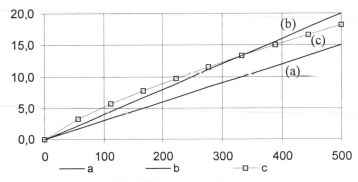

Figure 8. Fig. 8. Curve of flow $\tau(\gamma)$ according to Walburn-Schneck's model (c); average physiological values H = 45 ml / 100ml and TPMA = 2.5 g/ 100ml were assumed.

Curve of flow $\tau(\dot{\gamma})$ according to Walburn-Schneck's model; averagephysiological values $H = 45$ ml / 100ml and $TPMA = 2.5$ g/ 100ml were assumed.

4   Blood Flow in the Main Vessel

In considerations concerning the flow of some particular fluid, the question always appears: whether the assumption of the continuity of the medium is well founded. If we assume the dimensions of a blood cell d_c and vessel d_v as parameters [14]:

$$\Lambda_{c,V} = \frac{d_c}{d_V} \tag{17}$$

then assuming the data in section 2, referring to vessel dimensions (Fig. 1) and blood cells (Fig. 3), three different circulation categories can be distinguished:

- the capillary circulation comprises approximately $1.2 \cdot 10^{+9}$ vessels, usually not longer than 1mm, where $\Lambda_{c,v} \leq 1$;
- the micro- circulation consists of a vast network ($1.0 \cdot 10^{+8}$) of vessels with lengths in the range of 2 ... 10 mm, for which we have $\Lambda_{c,v} \approx 3 \ldots 50$;
- the main circulation comprises several thousand vessels of the largest diameters and lengths 10 ... 40 mm and more, for which we have $\Lambda_{c,v} > 50$. Every one of the above-mentioned categories demands a proper approach to the construction of a phenomenological blood flow model.

In the capillary circulation the blood flow should be modelled as a phenomenon of displacement of a deformable body (blood cells) due to the plasma thrust, generated by pressure differences and flexible walls of the vessels. Vessels, which are considered as belonging to microcirculation, have a slightly larger diameter; therefore the blood cell does not contact the wall directly. Due to the occurrence of high lateral (transverse) gradients, the morphological components gather along the channel axis separated from the wall by a thin layer of plasma. Eventually, blood movement is a superposition of the movements of two different (especially in the rheological aspect) media. For both of them separately an assumption about their continuity is permissible. But concerning the main circulation this assumption is hardly quite reasonable. Despite the fact that even in the last case changes of the contamination of blood cells in their cross-section do occur with an increasing tendency towards the axis, the differences are lower than in microcirculation.

The phenomenon of blood flow in the main vessels ($\Lambda_{c,v} > 50$) can be presented with satisfactory (good) precision, by a system of equations for mass and momentum balances and the momentum of the real medium [3,4,5]. Due to the axial symmetry of blood flow through a vessel, it is better to express the equations of motion in cylindrical coordinates; thus the balance of mass can be expressed as:

$$\frac{1}{r}\frac{\partial}{\partial r}ru_r + \frac{1}{r}\frac{\partial u_\theta}{\partial \theta} + \frac{\partial u_z}{\partial z} = 0 \tag{18}$$

whereas the equation of the momentum balance (along the z axis) can be written as follows:

$$\rho\left(\frac{\partial u_z}{\partial t} + u_r\frac{\partial u_z}{\partial r} + \frac{r_\theta}{r}\frac{\partial u_z}{\partial \theta} + u_z\frac{\partial u_z}{\partial z}\right) = -\frac{\partial p}{\partial z} +$$
$$\left(\frac{1}{r}\frac{\partial}{\partial r}r\tau_{rz} + \frac{1}{r}\frac{\partial \tau_{\theta z}}{\partial \theta} + \frac{\partial \tau_{zz}}{\partial z}\right) \tag{19}$$

where the respective stress coordinates are expressed by the following formulas:

$$\tau_{rz} = \eta_a\left(\frac{\partial u_z}{\partial r} + \frac{\partial u_r}{\partial z}\right) \tag{20}$$

$$\tau_{\theta r} = \eta_a\left(r\frac{\partial}{\partial r}\frac{u_\theta}{r} + \frac{1}{r}\frac{\partial u_r}{\partial \theta}\right) \tag{21}$$

$$\tau_{zz} = \eta_a\frac{\partial u_z}{\partial z} \tag{22}$$

For the case of flow through a vessel, the longitudinal dimension l of the canal is longer than the dimension of the cross-section d, i.e. $l >> d$, and therefore the axial component of transport is several times greater than the radial and tangent components. Therefore

$$O\left(\frac{u_\theta}{u_z}\right) = O\left(\frac{u_r}{u_z}\right) = 0 \tag{23}$$

the partial derivatives

$$O\left(\frac{\partial u_r}{\partial r}\right) = O\left(\frac{\partial u_\theta}{\partial \theta}\right) = ... = 0, \tag{24}$$

and at the same time

$$\tau_{rz} = \eta_a\frac{\partial u_z}{\partial r} >> O(\tau_{\theta r}) \approx O(\tau_{zz}). \tag{25}$$

Assuming the model of blood rheology according to equation (16) and assuming a stable flow, we finally obtain:

$$\frac{dp}{dz} = \frac{1}{r}\frac{d}{dr}\left(rc_1 e^{c_2H + c_4\frac{TPMA}{H^2}}\frac{du_z}{dr}^{1-c_3H}\right) \tag{26}$$

It is an ordinary differential equation, for which an integral in reference to dr can be easily obtained. Taking into account the adhesion of fluid on a vessel wall $(u_z|_{r=ro} = 0)$, we obtain an equation for velocity distribution in the cross-section:

$$u_z = -\frac{1-c_3H}{2-c_3H}\left(2c_1e^{c_2H+c_4\frac{TPMA}{H^2}}\right)^{\frac{1}{c_3H-1}}\frac{dp}{dz}^{\frac{1}{1-c_3H}}r_0^{\frac{2-c_3H}{1-c_3H}}\left(1-\frac{r}{r_0}\right)^{\frac{2-c_3H}{1-c_3H}} \tag{27}$$

The blood volume flux passing through a vessel, is given by the formula:

$$\dot{V} = \int_0^{r_0} 2\pi u_z r\,dr \tag{28}$$

substituting u_z from the formula (27) we obtain:

$$\dot{V} = -2\pi\frac{1-c_3H}{2-c_3H}\left(2c_1e^{c_2H+c_4\frac{TPMA}{H^2}}\right)^{\frac{1}{c_3H-1}}\frac{dp}{dz}^{\frac{1}{1-c_3H}}r_0^{\frac{2-c_3H}{1-c_3H}}\int_0^{r_0}\left(1-\frac{r}{r_0}\right)^{\frac{2-c_3H}{1-c_3H}}r\,dr \tag{29}$$

and after integrating, we obtain the form:

$$\dot{V} = -\pi\frac{1-c_3H}{4-3c_3H}\left(2c_1e^{c_2H+c_4\frac{TPMA}{H^2}}\right)^{\frac{1}{c_3H-1}}r_0^{\frac{4-3c_3H}{1-c_3H}}\frac{dp}{dz}^{\frac{1}{1-c_3H}} \tag{30}$$

Determining the average velocity in the cross-section in the traditional way as $u_0 = \dot{V}/\pi r_0^2$, and using formula (30) we obtain the next relationship:

$$u_0 = -\frac{1-c_3H}{4-3c_3H}\left(2c_1e^{c_2H+c_4\frac{TPMA}{H^2}}\right)^{\frac{1}{c_3H-1}}r_0^{\frac{4-3c_3H}{1-c_3H}}\frac{dp}{dz}^{\frac{1}{1-c_3H}} \tag{31}$$

Figure 9 presents the related velocity distribution in the cross-section of the blood vessel. It has been calculated basing on equation (31) for three values of the hematocrit number $H = 30, 45, 60$ ml/ 100ml. Additionally, the profile of velocities for the Newtonian model is also show.

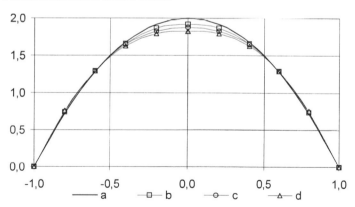

Figure 9. Velocity distribution u_z / u_0 in the vessel cross-section in the Newtonian model (a) and non-Newtonian model when $H = 30$ (b), 45 (c) and 60 ml/ 100ml (d).

The velocity profile takes a characteristic parabolic shape, but it is not a second-order parabola. Exponent values are slightly greater than 2 for the average level of hematocrit $H = 40.0 ... 45.0$ ml / 100ml i.e. $2.25 ... 2.29$, thus usually approximately 15% more than in the case of the Newtonian fluid. This increase is connected with the hematocrit number (see equation (32)). It is confirmed by more gradually homogenous velocity profiles and due to this, the average velocity is closer to the axial velocity, and the profile is slightly flatter.

Taking into account equation (27) for the velocity distribution in the cross-section and assuming that the maximum velocity u_{max} in a particular cross-section is equal to the axial velocity (for $r = 0$); it can be easily calculated:

$$u_{max} = -\frac{1-c_3 H}{2-c_3 H}\left(2c_1 e^{c_2 H + c_4 \frac{TPMA}{H^2}}\right)^{\frac{1}{c_3 H - 1}} \frac{dp}{dz}^{\frac{1}{1-c_3 H}} r_0^{\frac{2-c_3 H}{1-c_3 H}} \tag{32}$$

and the ratio of both these characteristic velocities (u_0 and u_z) is described by the following equation:

$$\frac{u_0}{u_{max}} = \frac{2-c_3 H}{4-3c_3 H} \tag{33}$$

The dependence on the hematocrit number (34) is shown in figure 10.

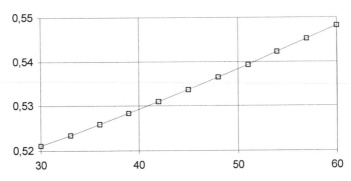

Figure 10. The ratio of the average to the maximum velocity u_0 / u_{max} against the hematocrit number H.

In the case of blood flow the average velocity reaches more than 50% of the maximum velocity in a particular cross-section. As the pathology develops (connected with an increase of the hematocrit number), this proportion gets higher and higher.

One of the more important practical applications of Hagen-Poiseuille's model is the possibility of determining analytically the pressure drop. After a suitable transformation of equation (31) we obtain:

$$-\left(\frac{1-c_3 H}{4-3c_3 H}\right)^{1-c_3 H}\left(2c_1 e^{c_2 H + c_4 \frac{TPMA}{H^2}}\right)^{-1} r_0^{1-c_3 H} \frac{dp}{dz} = u_0^{1-c_3 H}. \tag{34}$$

After integration along the canal with the length l_0 between the two arbitrarily chosen sections 1-1 and 2-2, we obtain the following formula:

$$-\frac{1-c_3H}{4-3c_3H}^{1-c_3H}\left(2c_1e^{c_2H+c_4\frac{TPMA}{H^2}}\right)^{-1}r_0^{1-c_3H}\left(p_2-p_1\right)=u_0^{1-c_3H}l_0 \tag{35}$$

Considering the pressure drop variation between the initial section 1-1 and the final section 2-2 as:

$$\Delta p = p_2 - p_1 \tag{36}$$

we obtain the formula:

$$\Delta p = 2c_1e^{c_2H+c_4\frac{TPMA}{H^2}}\left(\frac{4-3c_3H}{1-c_3H}\right)^{1-c_3H}\frac{l_0}{r_0^{2-c_3H}}u_0^{1-c_3H} \tag{37}$$

Similarly, as in the case of Dracy-Weisbach, the pressure drop can be characterised by the hydrodynamic friction coefficient (coefficient of linear losses):

$$\lambda = \frac{\Delta p\dfrac{d_0}{l_0}}{\dfrac{\rho u_0}{2}} \tag{38}$$

Therefore finally, based on equation (38), we have the dependence for the coefficient λ:

$$\lambda = 2^{4-c_3H}\frac{c_1}{\rho u_0^{1-c_3H}d_0^{1-c_3H}}e^{c_2H+c_4\frac{TPMA}{H^2}}\left(\frac{4-3c_3H}{1-c_3H}\right)^{1-c_3H} \tag{39}$$

In figure 11, the hydrodynamic friction coefficient is presented as a function of the average flow velocity of blood in the vessel (calculated by means of formula (40)). Calculating the values of this coefficient, it was assumed that the blood density $\rho = 1055$ kg/m^3, the protein level $TPMA = 2.5$ g/ 100ml and the vessel diameter, through which the flow takes place, $d_0 = 10$mm. The figure also show the coefficient λ for the Newtonian model for which:

$$\lambda = \frac{64}{Re} \tag{40}$$

assuming a value of dynamic viscosity coefficient $\eta = 3.5$cP.

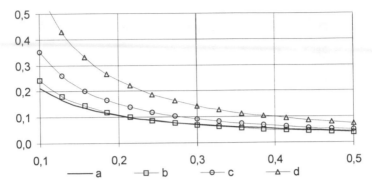

Figure 11. Distribution of the hydrodynamic friction coefficient λ in the Newtonian model (a) and for non-Newtonian model if H=30 (b), 45 (c) and 60 (d) ml/ 100ml.

The coefficient λ reaches the highest values in the range of low average velocities and gradually drops as the velocity increases. It is an obvious consequence of the decreasing influence of the viscosity factors in favour of inertial ones. In the range of average values of the hematocritic number $H = 30 \ldots 60$ ml / 100ml, in the case of a velocity of 0.1 m/s, it varies in the range $0.241 \ldots 0.593$, but when the velocity has reached 0.5m/s, it lowers to the range $0.039 \ldots 0.073$ (in the same circumstances for the Newtonian model it is 0.212 and 0.042, respectively). The values of the coefficient λ depend essentially on the hematocritic number, which can be easily seen in figure 11. Even in the case of hematocrit values, within the physiological standard (it changes two and a half times).

Figure 12 shows the influence of globulines and fibrinogenes levels on the value of the coefficient (keeping up a constant value of the hematocrit number $H =$ 45 ml /100ml).

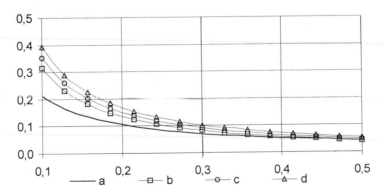

Figure 12. Distribution of the hydrodynamic friction coefficient λ in the Newtonian model (a) and non-Newtonian model in the case when $TPMA=1.0$ (b), 2.5 (c) and 4.0 (d) g/ 100ml.

The influence of the protein level, measured by *TPMA*, is evidently weaker. Assuming a variation of *TPMA* in the range 1.0 ... 4.0 g/100 ml, in the case of the average velocity 0.1 m/s, it varies from 0.316 ... 0.393, but for the average velocity 0.5 m/s (the interval is as follows: 0.044 ... 0.054). It means that deviations less than 20% occur when for the case of hematocryte these deviations is up to 50%.

Finally, at the end of the influence analysis of flow parameters, it should be added that a particular important and general statement can be established. For example, let's consider the equation (31) describing the velocity profile in the cross-section of a vessel in relation to the average velocity and let us investigate these ratio as the hematocritic numbers decreases:

$$\lim_{H \to 0} \frac{u_z}{u_0} = \lim_{H \to 0} \frac{4 - 3c_3 H}{2 - c_3 H} \left(1 - \left(\frac{r}{r_0}\right)^{\frac{2 - c_3 H}{1 - c_3 H}}\right) = 2\left(1 - \left(\frac{r}{r_0}\right)^2\right) \tag{41}$$

and simultaneously, according to the equation (34)

$$\lim_{H \to 0} \frac{u_0}{u_{max}} = \lim_{H \to 0} \frac{2 - c_3 H}{4 - 3c_3 H} = \frac{1}{2} \tag{42}$$

which gives us in the well-known result of the Hagen-Poiseuille's flow for the Newtonian fluids. Therefore, it may be said that the Newtonian model is a particular boundary state of the blood flow in a vessel, the more adequate the smaller the hematocritic number.

5   Sensitivity of Flow Parameters to Changes of the Blood State

If we try to analyse the influence of blood state parameters, first and foremost a method should be chosen allowing to asses this state in a simple and effective way. In the present paper, the classic concept of Bode's sensitivity has been applied. It was derived from the control theory and recently it has also been used in biomechanics [1]. And so, if a particular quantity v depends on the parameter x, its sensitivity is defined in the following way:

$$S_{v,x} \overset{def}{=} \lim_{\Delta x \to 0} \frac{\frac{\Delta v}{v}}{\frac{\Delta x}{x}} \tag{43}$$

It is easy to notice that in the light of the assumed definition (for example) if $v \sim x^n$, then the sensitivity $S_{v,x} = n$, and the constant function $S_{v,x} = 0$.

First, we consider the sensitivity of the velocity field to changes of the hematocritic number. Calculations were performed in this case for the non-dimensional distribution of the ratio u_z / u_0 described by formula (32); the results are presented in figure 13.

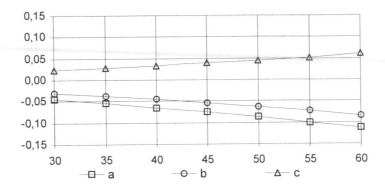

Figure 13. Sensitivity of velocity distribution u_x / u_o on changes of the hematocrit number for $r / r_o = 0.00$ (a), 0.25 (b) and 0.75 (c).

The sensitivity of the distribution of the velocity field to changes of the hematocritic number (the only parameter influencing the velocity profile) is not too extensive because the considered range $H = 30 \ldots$ ml/ 100ml varies in the range - 0.11 ... 0.06, but it increases together with the increase of the hematocrit number. In the neighbourhood of the wall, the sensitivity is slightly higher than zero ($r/r_o = 0.75$ - the range is 0.02 ... 0.06), thus if the hematocrit increases, the velocities in that area increase slightly. Moving towards the canal axis the sensitivity decreases and reaches a negative extremum for $r/r_o = 0.0$, which is in the range -0.05 ... - 0.11; the velocity decreases as the hematocrit increases.   The result of the presented behaviour of the velocity field sensitivity (Fig. 9) is a tendency to make the distribution of velocities increasingly flutter i.e. to decrease the velocities farther away from the walls and to increase them near the walls with the increasing values of the hematocritic number.

The sensitivity of the hydrodynamic friction coefficient was calculated for the initial part of the aorta, assuming an ejection volume equal to 100 cm³, the period of ejection *0.8s* and the diameter 25mm. The remaining parameters were assumed in the same way as in the calculations presented in figure 11. Figure 14 shows the sensitivity of the coefficient λ with the hematocrit number. Changes of the hematocritic number exert an essential influence on the coefficient λ. Its sensitivity for $H = 60$ ml /100ml reaches 2.35 ... 2. 07. In the range of lower hematocrits this influence is lower and lower and for example for $H = 30$ ml /100ml it reaches -0.1 ... 0.87. Additionally, the sensitivity $S_{\lambda, H}$ is modified by the influence of the protein level *TPMA*, so that the greater the *TPMA* the lower is the sensitivity; on the other hand, the influence of protein level decreases as the hematocrit increase. For example for $H = 30$ ml /100ml *TPMA* = 2.5 g/ml, disturbances of ± 1.5 g/100ml in the *TPMA* cause a change in the sensitivity of more than 120%, but for $H = 60$ ml /100ml this changes is only 5%.

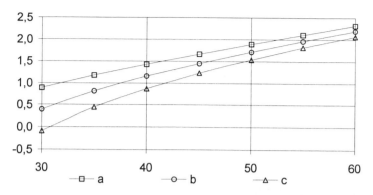

Figure 14. Sensitivity of the hematocrit friction coefficient λ to changes of the hematocrit for *TPMA* = *1.0* (a); *2.5* (b) and *4.0* (c) *g/ml.*

Figure 15, presents the sensitivity of the hydrodynamic friction coefficient, λ, to changes of the protein level *TPMA*, (i.e. globulins and fibrogen). The range of *TPMA* concentration is 1.0 ... 4.0g/ml i.e. it can be considered to be as physiologically acceptable.

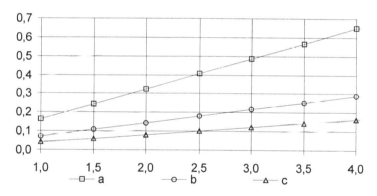

Figure 15. Sensitivity of the hematocritic friction coefficient λ to changes of the protein level for $H = 30$ (a); *45* (b) and 60 (c) *ml /100ml.*

Within the considered range of variability, the sensitivity is greater than zero and it increases linearly against the protein level *TPMA*, but it is modified by the hematocritic number. For $H = 30$ ml/ 100ml, it was obtained that $S_{\lambda\ TPMA} = 0.16$ ... 0.65 and for $H = 60$ ml/ 100ml just four times less (0.04 ... 0.16). The increase of the hematocritic number causes a general effect of reducing the sensitivity so that at $H = 75$ ml/ 100ml it does not exceed 0.1.

It should be stressed that the given results are restricted to blood parameter values whose maxima do not exceed physiological standard i.e. $H \leq 60$ ml/ 100ml and *TPMA* ≤ 4.0 g/ 100ml. Let's consider now the case when *TPMA* = 2.5 g/ 100ml, and H = 79 ml/ 100ml. Then, the sensitivity of the hydrodynamic friction coefficient whit respect to the hematocritic number is greater than 3, therefore pressure drops (energy) will increase proportionally to the exponent 3. Thus in the case of pathological states, the sensitivity will raise and raise so that the blood parameters are of great importance also with reference to the flow behaviour.

6   Final Remarks

Searching for solutions of the problem of blood flow through the main vessel considered above can be achieved using different methods; besides the presented analytical approach, experimental and numerical investigations are possible (as for example in the paper [20]). Nevertheless, it should be stressed that the assumed method displays the following advantages:

- contrary to experiments, expensive and unique laboratory equipment need not be used (due to the exceptional problem and atypical medium);
- the application of Hagen-Poiseuille's flow model permits analytical   solutions, e.g. precise ones in the class of the applied model;
- the geometric simplicity of the model and knowledge of solutions concerning the problem of Newtonian fluid provides a better relationship between hematology and the physical side of this phenomenon.

The performed detailed analysis shows the following regularities connected with hematological and flow parameters:

- a strong influence of pseudo-plastic properties; due to the decreasing shear velocity, modified by values of the hematocrit number and levels of fibrinogene and globulines;
- a decrease of the maximum velocity in relation to the average velocity, which is confirmed by the flatter profile of velocities together with an increase of the hematocritic number and protein level measured by the *TPMA* indicator;
- large sensitivity of the considered parameters to the hematocritic number and small sensitivity to changes of the levels of fibrinogene and globulines.

The influence of blood state parameters, visible by the rheological properties of the blood, causes that the obtained results differ greatly from those achieved on the Newtonian model. Non-Newtonian effects and their consequences and more important when the blood composition differs from the proper (regular) one in the human body which indicates a particular correlation between the pathological states of the blood-flow system.

References

1. McDonald D.A., *Blood flow in arteries*, (Camelot Press, Southampton, 1974).
2. Milnor W.R., *Hemodynamics*, (Williams&Wilkins, Baltimor 1989).
3. Wormersly J.R., Method of the calculation of velocity, rate of flow and viscous drag in arteries when the pressure gradient is known, *J. Physiol.* **127** (1955) pp. 553-563.
4. Wormersly J.R., Oscillatory motion of a viscous liquid in a thin-walled elastic tubes, *Phil. Mag.* **46** (1955) pp. 191-221.
5. Rodkiewicz C.M., Sinha P., Kennedy J.S., On the application of constitutive equation for whole human blood, *Journal of Biomechanical Engineering* **112** (1990) pp. 198-206.
6. Ganong W.F., *Fizjologia. Podstawy fizjologii lekarskiej*, (Państwowy Zakład Wydawnictw Lekarskich, Warszawa 1994).
7. Hope R.A., Longmore J.M., Moss P.A.H., *Oxford Handbook of Clinical Medicine*, (Oxford University Press, Oxford 1989) pp. 84-102.
8. Miętkiewski E., *Zarys fizjologii lekarskiej*, (Państwowy Zakład Wydawnictw Lekarskich, Warszawa 1984).
9. Chmiel H., Walitza E., *On the rheology of blood and synovial fluid*, (Research Studies Press, New York 1980).
10. Bayliss L.E., Rheology of blood and lymph, In *Deformation and flow in biological systems*, ed. by Frey-Wissling A., (North Holland Publ., Amsterdam 1952) pp. 355-415.
11. Orzechowski Z., *Przepływy dwufazowe. Jednowymiarowe i adiabatyczne*, (P W N. Warszawa 1990).
12. Dintenfass L., Thixotropy of blood at very low rates of shear, *Kolloidzeitschrift* **180** (1961) p. 160.
13. Chien S., Shear dependence of effective cell volume as a determinant of blood viscosity, *Science* **168** (1970) pp. 977-979.
14. Quemada D., Blood rheology and its implication in flow of blood. In *Arteries and arterial blood flow*, ed. by Rodkiewicz C.M., (Springer Verlag, New York 1983) pp. 1-127.
15. Scott Blair G.W., An equation for the flow of blood, plasma and serum, through glass capillaries, *Nature* (1959) pp. 613-615.
16. Casson N., A flow equation for pigment-oil suspension of printing ink type, In *Rheology of Disperse Systems*, ed. by Mill C.C., ed., (Pergamon Press, Oxford 1959) pp. 84-102.
17. Merrill E.W., Rheology of blood, *Physiol. Rev.* **4** (1969). pp. 863-888.
18. Eilers R., Ritter Chr., Rau G., Influence of non-Newtonian blood properties on numerical flow simulation at mechanical heart valve prostheses, In *Biofluid Mechanics*, ed. by Liepsh D., (VDI Verlag, Munich 1994) pp. 693-700.
19. Walburn F.J., Schneck D.J., A constitutive equation for whole human blood, *Biorheology* **13** (1976) pp. 201-210.

20. Wojnarowski J., Mirota K., Własności reologiczne krwi ludzkiej w modelowaniu opływu sztucznej zastawki (*XXX Sympozjon Modelowanie w Mechanice*, Wisła-Gliwice 1996) pp. 251-256.
21. Hoppensteadt C.F., Peskin C.S., *Mathematics in medicine and the life sciences*, (Springer Verlag, New York 1992).

THERMOMECHANICS OF BONE-IMPLANT INTERFACES AFTER CEMENTED JOINT ARTHROPLASTY : THEORETICAL MODELS AND COMPUTATIONAL ASPECTS

LALAONIRINA R. RAKOTOMANANA

Institut de Recherche Mathématiques de Rennes
UMR 6625 CNRS - Université de Rennes 1, 35 042 Rennes, France
E-mail: lalaonirina.rakotomanana@univ-rennes1.fr

NIRINA A. RAMANIRAKA

Centre de Recherche Orthopédique, EPF Lausanne, Suisse

Numerous factors are known to influence the fixation of orthopedic implants, such as the bone quality, the implant design and the bone-implant interface. Accurate evaluation of the bone-implant micromotion and interfacial stress is a keystone for determining the mechanical environment acting on the surrounding bone tissue and cells. This work presents some theoretical models and computational methods to investigate the initial fixation of orthopedic implants. Both mechanical and thermal aspects are considered. A cemented femoral component was chosen to illustrate the developed models.

1 Introduction

After joint arthroplasty, failure of cemented implants is commonly related to wear particles resulting from cement fracture and fragmentation due to high stress and enhanced by a migration of debris particles along cement-bone interface. More specifically, after cemented Total Hip Replacement, failures at stem-cement interface and at bone-cement interface mainly resulted from the occurrence of abnormally high cement stresses and excessive micromotions [1,2]. High compressive stress was source of cement fracture and enhanced the stem subsidence while excessive slipping enhanced the creation of cement debris. Bone-cement slipping leaded to a necrosis of bone that interdigitize with the cement. Interfaces roughness strongly influenced the stress and micromotions. Improving the cement-stem bonding increased stress at cement-bone interface [2]. The cement thickness was another factor influencing sensibly stress magnitude, observed either clinically or experimentally [3]. Mechanical and thermal problems at the interface are very sensitive to minor changes of contacting boundaries of the bone, cement and implants. This sensitivity is increased by thermoelastic distortion [4]. It may have important consequence of the long term reliability of the implant anchorage, particularly on the generation of microcraks within the cement mantle. Some biological complications might be related to the heat generated by the exothermic polymerization of the orthopaedic cement. Bone necrosis at the bone-cement interface may occur during and after the surgical implantation. This may induce bone resorption at the bone-cement interface, leading implant lose. A brief review allows to cite some variables which may affect the interface behaviour: micromotions and stresses, wear and debris at the bone-cement and cement-stem interfaces, bone remodeling and formation of fibrous tissue, and thermal effects due to cement and to frictional contact at the interface.

Computational techniques are more and more used to calculate these mechanical and thermal variables trying to simulate clinical situations [5]. However, coupling of all of these variables drastically increases the problem difficulty. To name but several, stem temperature, cement temperature, polymerization temperature, thermoelastic properties of bone, cement and stem including the shrinkage stresses of the cement [6], certainly affect the quality of the anchorage either immediatly after operation or in the long term. As a starting point, accurate calculation of the bone-implant relative micromotions and interfacial stresses becomes a keypoint for any quantitative analysis of the mechanical environment acting on the bone tissue and cells. Some numerical algorithms are also very sensitive to solution parameters [7] and it is always reasonable to question about exactness of theoretical models and efficiency of numerical computations.

The purpose of this work was to present theoretical models and computational methods to analyse the fixation of orthopedic implants. Only some thermomechanics aspects are considered. This paper focuses on the development of theoretical models for interface thermomechanics, and of numerical algorithms with application to cemented arthroplasty.

2   Thermomechanics of the Interface

2.1   *Interface thermokinematics*

Interface kinematics includes cohesion-decohesion and adherence and sliding between the contacting surfaces. We limit to infinitesimal deformation within body and infinitesimal relative displacement at the interface. Considering two deformable bodies B and B'. B is limited by the boundary $\partial B = \partial B_p \cup \partial B_c \cup \partial B_u$, where on ∂B_u displacement is imposed, on ∂B_p stress vector is imposed, and ∂B_c is the contact surface between B and B'. The same decomposition holds for B' (see Fig. 1). B' is called the target and B the striker. For any point $M \in \partial B_c$, the proximal point $M^\perp$ the position of which

$\mathbf{OM}^{\perp} = \varphi'(M^{\perp}, t)$ is the nearest of that to M [8]:

$$\mathbf{OM}^{\perp} = \operatorname{argmin} \left[\frac{1}{2} \|\varphi(M, t) - \varphi'(M', t)\|^2 \right]_{M' \in \partial B'_c} \tag{1}$$

The gap vector $\mathbf{d}$ is defined by [8] $\mathbf{d} \equiv \varphi(M, t) - \varphi'(M^{\perp}, t) = \mathbf{M}^{\perp}\mathbf{M}$ where $\mathbf{OM}^{\perp}(M, t)$. The normal gap vector d_n is the component of $\mathbf{d}$ along $\mathbf{n}'(M^{\perp}, t)$, normal at $M^{\perp}$ (normal micromotions):

$$d_n(M, t) \equiv \mathbf{n}'(M^{\perp}, t) \cdot [\varphi(M, t) - \varphi'(M^{\perp}, t)] \tag{2}$$

Three situations may occur: $d_n > 0$ gap, $d_n = 0$ contact and $d_n < 0$ penetration. The slipping vector (shear micromotions) is then given by:

$$\mathbf{d}_T(M, t) \equiv \mathbf{d}(M, t) - d_n(M, t)\mathbf{n}'(M^{\perp}, t) \tag{3}$$

The rate of the gap vector is the relative velocity of M located in $\varphi(M, t) \in \partial B_c$, with respect to $M^{\perp}$, located in $\varphi'(M^{\perp}, t) \in \partial B'_c$. :

$$\dot{\mathbf{d}} = \frac{d}{dt}\varphi(M, t) - \frac{d}{dt}\varphi'(M^{\perp}, t) \tag{4}$$

It is not frame indifferent unless the normal gap vanishes. However, for small relative displacements, there is no need to introduce objective relative velocity. Normal and tangential projections of $\dot{\mathbf{d}}$ are respectively:

$$\dot{\mathbf{d}}_n = \left[\mathbf{n}'(M^{\perp}, t) \otimes \mathbf{n}'(M^{\perp}, t) \right] \dot{\mathbf{d}}(M, M^{\perp}, t)$$
$$\dot{\mathbf{d}}_T = \left[\mathbb{I} - \mathbf{n}'(M^{\perp}, t) \otimes \mathbf{n}'(M^{\perp}, t) \right] \dot{\mathbf{d}}(M, M^{\perp}, t) \tag{5}$$

By analogy with $\mathbf{d}$, we define the temperature-gap $\delta\theta$ and its rate:

$$\delta\theta = \delta\theta(M, t) = \theta(M, t) - \theta'(M^{\perp}(M, t), t) \qquad \dot{\delta\theta} = \frac{d\delta\theta}{dt} \tag{6}$$

2.2 Conservation laws at the interface

Interface thermomechanics can be described by the position $\mathbf{OM}(t)$ and contact gap $\mathbf{d}(M, t)$; the absolute temperature $\theta(M, t)$ and temperature gap $\delta\theta(M, t)$; the contact stress vectors $\mathbf{p}(M, t)$ and $\mathbf{p}'(M', t)$; the entropy $s(M, t)$ and interfacial entropy $s_c(M, t)$; the internal energy $e(M, t)$ and interfacial energy $e_c(M, t)$; the contact heat fluxes $q(M, t)$, $q'(M', t)$; the body force $\rho b(M, t)$ and volume heat source $r(M, t)$ [9].

Let $\mathcal{A}$ be the quantity which varies within $\Sigma = B \cup B'$ and $\mathcal{B}$ the volume production and $\mathcal{C}(\partial\Sigma, t)$ the flux exchange with the external environment. The generic conservation law holds:

$$\frac{d}{dt}\mathcal{A}(\Sigma, t) \gtreqless \mathcal{B}(\Sigma, t) + \mathcal{C}(\partial\Sigma, t). \tag{7}$$

Any subsystem of Σ must satisfy (7). When B and B' shrink down respectively to ∂B_c and $\partial B'_c$, their volume vanishes while the areas of the contact surfaces remain finite. Two situations may occur: (a) *no contact* the target surface and the striker surface are empty sets, (b) *contact* there is a bijective map between the two contacting surfaces.

1. **For linear momentum.** We introduce:

$$A(\Sigma, t) = \int_B \rho \mathbf{v} dv + \int_{B'} \rho' \mathbf{v}' dv' \tag{8}$$

$$B(\Sigma, t) = \int_B \rho \mathbf{b} dv + \int_{B'} \rho' \mathbf{b}' dv' \tag{9}$$

$$C(\partial \Sigma, t) = \int_{\partial B} \mathbf{p}_n da + \int_{\partial B'} \mathbf{p}'_n da' \tag{10}$$

Since this relation must be satisfied for any subsystem of Σ, we obtain the three equations for any point of B, B', and $\partial B_c \cap \partial B'_c$ respectively:

$$\rho \dot{\mathbf{v}}(M, t) = \rho \mathbf{b}(M, t) + \mathrm{div}\sigma(M, t)$$
$$\rho' \dot{\mathbf{v}}'(M', t) = \rho' \mathbf{b}'(M', t) + \mathrm{div}\sigma'(M', t)$$
$$\mathbf{p}(M, t) + \mathbf{p}'(M^\perp(M, t), t) = 0 \tag{11}$$

2. **For angular momentum.** Consider the angular momentum with respect to any fixed point O. We introduce:

$$A(\Sigma, t) = \int_B \mathbf{OM} \times \rho \mathbf{v} dv + \int_{B'} \mathbf{OM}' \times \rho' \mathbf{v}' dv' \tag{12}$$

$$B(\Sigma, t) = \int_B \mathbf{OM} \times \rho \mathbf{b} dv + \int_{B'} \mathbf{OM}' \times \rho' \mathbf{b}' dv' \tag{13}$$

$$C(\partial \Sigma, t) = \int_{\partial B} \mathbf{OM} \times \mathbf{p} da + \; + \int_{\partial B'} \mathbf{OM}' \times \mathbf{p}' da' \tag{14}$$

Application of this general relation for any subsystem of Σ induces the three equations for any point of B, B', and $\partial B_c \cap \partial B'_c$ respectively:

$$\sigma(M, t) = \sigma^T(M, t) \quad \sigma'(M', t) = \sigma'^T(M', t)$$
$$\mathbf{MM}^\perp(M, t) \times \mathbf{p} = 0 \tag{15}$$

3. **For internal energy.** For the energy, $A(\Sigma, t)$ includes the kinetic $K(\Sigma, t)$ and internal energy $\mathcal{E}(\Sigma, t)$. This later is not additive:

$$\dot{\mathcal{E}}(\Sigma, t) = \dot{\mathcal{E}}(B, t) + \dot{\mathcal{E}}'(B', t) + \int_{\partial B_c} \frac{de_c}{dt} da \tag{16}$$

in which the third term of the right side represents the interfacial energy between B and B'. We thus define the kinetic energy, internal energy, power of external forces, and volume and heat fluxes respectively:

$$\mathcal{K}(\Sigma, t) = \frac{1}{2} \int_B \rho v^2 dv + \int_{B'} + \frac{1}{2} \int_{B'} \rho' \mathbf{v}'^2 dv' \tag{17}$$

$$\mathcal{E}(\Sigma, t) = \int_B \rho e dv + \int_{B'} \rho' e' dv' + \mathcal{E}(B \cap B', t) \tag{18}$$

$$\mathcal{P}_e(\Sigma, t) = \int_B \rho \mathbf{b} \cdot \mathbf{v} dv + \int_{\partial B} \mathbf{p} \cdot \mathbf{v} da$$

$$+ \int_{B'} \rho' \mathbf{b}' \cdot \mathbf{v}' dv' \int_{\partial B'} \mathbf{p}' \cdot \mathbf{v}' da' \tag{19}$$

$$\mathcal{Q}(\partial \Sigma, t) = \int_B r dv - \int_{\partial B} q da + \int_{B'} r' dv' - \int_{\partial B'} q' da' \tag{20}$$

The energy conservation law for ∂B_c and $\partial B'_c$ reduces to:

$$\int_{\partial B_c} \frac{de_c}{dt} da = \int_{\partial B_c} (\mathbf{p} \cdot \mathbf{v} - q) \, da + \int_{\partial B'_c} (\mathbf{p}' \cdot \mathbf{v}' - q') \, da' \tag{21}$$

Rigorously, $\partial B'_c$ may be split onto two parts $\partial B'_c - \partial B_c^\perp$ and $\partial B_c^\perp$. This allows us to transfer the calculation of the contact power and the heat flux on ∂B_c. Let us recall the contact power

$$\mathcal{P}_c(\partial B_c, t) \equiv \int_{\partial B_c} \mathbf{p} \cdot \mathbf{v} da + \int_{\partial B'_c} \mathbf{p}' \cdot \mathbf{v}' da' \tag{22}$$

The principle of Action-reaction, the fact that $\mathbf{p}' \equiv 0$ on $\partial B'_c - \partial B_c^\perp$, and the equilibrium of moment induce:

$$\mathcal{P}_c(\partial B_c, t) = \int_{\partial B_c} \mathbf{p} \cdot \dot{\mathbf{d}} da \tag{23}$$

The three local forms of the energy conservation hold:

$$\rho \dot{e}(M, t) = \sigma : \mathrm{grad}\mathbf{v}(M, t) + r(M, t) - \mathrm{div}\mathbf{J}_q(M, t)$$
$$\rho' \dot{e}'(M', t) = \sigma' : \mathrm{grad}\mathbf{v}'(M', t) + r'(M', t) - \mathrm{div}\mathbf{J}'_q(M', t)$$
$$\dot{e}_c(M, t) = \mathbf{p} \cdot \dot{\mathbf{d}}(M, t) - q(M, t) - q' \left(M^\perp(M, t), t \right) \tag{24}$$

4. **For entropy.** As for energy, we assume that the entropy $\mathcal{A} = \mathcal{S}$ is not additive and that there is interfacial entropy density:

$$\mathcal{S}(\Sigma, t) = \int_B \rho s dv + \int_{B'} \rho' s' dv' + \int_{\partial B_c} s_c da \tag{25}$$

The entropy supply includes volume and surface terms :

$$\mathcal{B}(\Sigma, t) = \int_B \frac{r}{\theta} dv + \int_{B'} \frac{r'}{\theta'} dv' \tag{26}$$

$$\mathcal{C}(\partial\Sigma, t) = -\int_{\partial B} \frac{q}{\theta} da - \int_{\partial B'} \frac{q'}{\theta'} da' \tag{27}$$

By applying the entropy inequality for any subsystem of Σ, we obtain for any point of B, B', and $\partial B_c \cap \partial B'_c$ respectively:

$$\rho \dot{s}(M, t) \geq \frac{r}{\theta}(M, t) - \mathrm{div}\left(\frac{\mathbf{J}_q}{\theta}\right)(M, t)$$

$$\rho' \dot{s}'(M', t) \geq \frac{r'}{\theta'}(M', t) - \mathrm{div}\left(\frac{\mathbf{J}'_q}{\theta'}\right)(M', t)$$

$$\dot{s}_c(M, t) \geq -\left(\frac{q}{\theta}(M, t) + \frac{q'}{\theta'}((M^\perp(M, t), t)\right) \tag{28}$$

2.3 Constitutive laws at the interface

Interfacial heat transfer is due to heat generation by friction and/or due to the temperature jump across the interface [10], requiring to take into consideration the principles of thermodynamics [11]. By assuming a rate dependence, generic constitutive functions reduce to [9]:

$$\mathfrak{I}_c = \hat{\mathfrak{I}}_c(d_n, \mathbf{d}_T, \theta, \delta\theta, \dot{d}_n, \dot{\mathbf{d}}_T, \dot{\theta}, \dot{\delta\theta}) \tag{29}$$

Definition 1: (Interfacial free energy) The interfacial free energy $\phi_c(M, t)$ distributed on ∂B_c is defined by $\phi_c \equiv e_c - \theta s_c$. We assume that ϕ_c is a (non differentiable) function of the normal contact gap d_n and the tangential contact gap $\mathbf{d}_T$ and a (differentiable) function of the temperature θ on ∂B_c, of the temperature-gap $\delta\theta$ and their rates.

Theorem 1: Let B and B' be two continua in contact with $\phi_c = \hat{\phi}_c(d_n, \mathbf{d}_T, \theta, \delta\theta, \dot{d}_n, \dot{\mathbf{d}}_T, \dot{\theta}, \dot{\delta\theta})$. Assume that the constitutive functions are not dependent on the second-order time derivatives of $(d_n, \mathbf{d}_T, \theta, \delta\theta)$, then:

$$\phi_c = \hat{\phi}_c(d_n, \mathbf{d}_T, \theta) \qquad q' \frac{\delta\theta}{\theta'} + \mathbf{J}_{\mathbf{d}_T} \cdot \dot{\mathbf{d}}_T \geq 0$$

$$\mathbf{J}_{\mathbf{d}_T} \equiv \mathbf{p}_T - \partial_{\mathbf{d}_T}\phi_c \qquad p_n \in \partial_{d_n}\phi_c \qquad s_c = -\frac{\partial\phi_c}{\partial\theta}$$

Proof. By substituting $\dot{s}_c$ from the above definition and $\dot{e}_c$ from (24), by introducing the temperature-gap, (28) reduces to [12,9]:

$$\mathbf{p} \cdot \dot{\mathbf{d}} - \dot{\phi}_c - \dot{\theta} s_c + \frac{q'}{\theta'}\delta\theta \geq 0$$

By introducing the rate of ϕ_c in the entropy inequality and by splitting the contact power into two parts (normal and tangential), we obtain:

$$(p_n - \partial_{d_n}\phi_c)\dot{d}_n + (\mathbf{p}_T - \partial_{\mathbf{d}_T}\phi_c) \cdot \dot{\mathbf{d}}_T - \left(s_c + \frac{\partial\phi_c}{\partial\theta}\right)\dot{\theta} + \frac{q'}{\theta'}\delta\theta$$

$$-\frac{\partial\phi_c}{\partial\delta\theta}\dot{\delta\theta} - (\partial_{\dot{d}_n}\phi_c)\ddot{d}_n - (\partial_{\dot{\mathbf{d}}_T}\phi_c)\ddot{\mathbf{d}}_T - \frac{\partial\phi_c}{\partial\dot{\theta}}\ddot{\theta} - \frac{\partial\phi_c}{\partial\dot{\delta\theta}}\ddot{\delta\theta} \geq 0 \qquad (30)$$

Following Coleman and Noll method [9] we consider constitutive functions independent of $\ddot{d}_n$, $\ddot{\mathbf{d}}_T$, $\ddot{\theta}$, and $\ddot{\delta\theta}$. We can arbitrarily choose second derivatives at the instant t^+. To satisfy (30), it is necessary for second order derivative coefficients to vanish:

$$\partial_{\dot{d}_n}\phi_c = 0 \qquad \partial_{\dot{\mathbf{d}}_T}\phi_c = 0 \qquad \frac{\partial\phi_c}{\partial\dot{\theta}} = 0 \qquad \frac{\partial\phi_c}{\partial\dot{\delta\theta}} = 0 \qquad (31)$$

Therefore ϕ_c and s_c, do not depend on either $\dot{\theta}$ or $\dot{\delta\theta}$. We can attribute arbitrary values for $\dot{\theta}$ and $\dot{\delta\theta}$ at t^+ without violating (30). It implies:

$$s_c = -\frac{\partial\phi_c}{\partial\theta} \qquad \frac{\partial\phi_c}{\partial\delta\theta} = 0 \qquad (32)$$

1. the interfacial free energy necessarily takes the form of:

$$\phi_c = \hat{\phi}_c(d_n, \mathbf{d}_T, \theta) \qquad (33)$$

2. the entropy inequality (30) reduces to:

$$q' \frac{\delta\theta}{\theta'} + J_{d_n}\dot{d}_n + \mathbf{J}_{\mathbf{d}_T} \cdot \dot{\mathbf{d}}_T \geq 0 \qquad (34)$$

in which:

$$J_{d_n} \equiv p_n - \partial_{d_n}\phi_c \qquad \mathbf{J}_{\mathbf{d}_T} \equiv \mathbf{p}_T - \partial_{\mathbf{d}_T}\phi_c \qquad (35)$$

For perfect unilateral contact, the normal relative velocity may take any value and that this must not violate (30). We can deduce:

$$p_n \in \partial_{d_n}\phi_c \qquad s_c = -\frac{\partial\phi_c}{\partial\theta} \qquad q' \frac{\delta\theta}{\theta'} + \mathbf{J}_{\mathbf{d}_T} \cdot \dot{\mathbf{d}}_T \geq 0 \qquad (36)$$

The first inclusion is obtained by noticing the inequality $\dot{d}_n \geq 0$. Indeed, we can always conclude that $p_n \in \partial_{d_n}\phi_c$, whatever form of ϕ_c is chosen. It is the indicator function of R_+ with respect to d_n. If the free energy does not depend on $\mathbf{d}_T$, we find the classical unilateral frictional laws [13]. To sum up, constitutive laws of the thermocontacts may be entirely reconstructed from free energy ϕ_c and quasi potential of dissipation ψ_c [11]:

$$\phi_c = \hat{\phi}_c(d_n, \mathbf{d}_T, \theta) \qquad \psi_c = \hat{\psi}_c\left(\dot{d}_n, \dot{\mathbf{d}}_T, \frac{\delta\theta}{\theta'}\right) \tag{37}$$

2.4   Coulomb dry friction

Usual laws for unilateral contact [13] may be included in the present framework. As a rule we start with local law (one point) and then extend the laws to discrete multidimensional contact, by defining the global displacement vector $\mathcal{U} \equiv (\mathbf{u}(M \in B), \mathbf{u}'(M' \in B'))$.

1. **Unilateral contact.** Unilateral contact laws are summarized by the classical Signorini relationships $d_n \geq 0$, $p_n \leq 0$, and $p_n\, d_n = 0$. Equivalently, multivalued contact law $p_n\,[d_n]$ and its inverse $d_n\,[p_n]$ may be derived from non differentiable two pseudo-potentials:

$$p_n \in \partial \mathbb{I}_{R_+}(d_n) \qquad d_n \in \partial \mathbb{I}_{R_-}(p_n) \tag{38}$$

where the interfacial free energy $\phi_c(d_n) = \mathbb{I}_{R_+}(d_n)$ is the indicator function of real numbers R_+ defined by:

$$\mathbb{I}_{R_+}(d_n) = \begin{cases} 0 & d_n \in R_+ \\ \infty & d_n \notin R_+ \end{cases} \tag{39}$$

$\partial \mathbb{I}_{R_+}(d_n)$ is the sub-differential of $\mathbb{I}_{R_+}(d_n)$ and $\mathbb{I}_{R_-}(p_n)$ its conjugate.

By considering (discrete) global gap vector as it applies in finite element analysis, the unilateral contact laws are (for p contact elements):

$$\mathcal{D}_n(\mathcal{U}) \in \mathbb{R}_-^p \qquad \mathcal{F}_n \in \mathbb{R}_+^p \qquad \mathcal{D}_n(\mathcal{U}) \cdot \mathcal{F}_n \tag{40}$$

These (in)-equations may be merged into the unilateral contact law [17]:

$$\mathcal{F}_n \in \partial \mathbb{I}_{\,\mathbb{R}_+^p}\,[\mathcal{D}_n(\mathcal{U})] \tag{41}$$

2. **Coulomb pure friction.** For pure friction, under a constant pressure $p_n = c^{ste}$, the dissipation potential is $\psi_c = \mathbb{I}_{C(p_n)}^*$. Two cases may occur: stick $\|\mathbf{p}_T\| + \mu p_n < 0$ and slip $\|\mathbf{p}_T\| + \mu p_n = 0$. These two situations may be expressed by means of a slip law, a Coulomb criterion and a complementary condition, μ being the friction coefficient:

$$\dot{\mathbf{d}}_T \|\mathbf{p}_T\| = \|\dot{\mathbf{d}}_T\| \mathbf{p}_T \quad \|\mathbf{p}_T\| + \mu p_n \leq 0 \quad \|\dot{\mathbf{d}}_T\|(\|\mathbf{p}_T\| + \mu p_n) = 0 \tag{42}$$

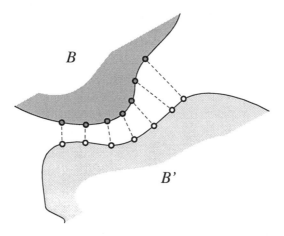

Figure 1. Global contact vectors. After discretization of two solids, the contacting surfaces ∂B_c and $\partial B'_c$ are discretized to define the contact elements (dashed lines). A global gap vector $\mathcal{D}_n$ whose components are the local contact gaps of each contact element $\mathcal{D}_n = \left(d_n^1, ..., d_n^p\right)$ where p is the number of contact elements. A global slipping vector is also defined $\mathcal{D}_T = \left(\mathbf{d}_T^1, ..., \mathbf{d}_T^p\right)$. Accordingly, global normal contact vector and shear contact vector are defined as $\mathcal{F}_n = \left(p_n^1, ..., p_n^p\right)$ and $\mathcal{F}_T = \left(\mathbf{p}_T^1, ..., \mathbf{p}_T^p\right)$ respectively.

It is worth to introduce conjugate potentials and their sub-gradients [13]:

$$\mathbf{p}_T \in \partial \mathbb{I}^*_{C(p_n)}\left(\dot{\mathbf{d}}_T\right) \qquad \dot{\mathbf{d}}_T \in \partial \mathbb{I}_{C(p_n)}(\mathbf{p}_T) \tag{43}$$

where $\mathbb{I}^*_{C(p_n)}\left(\dot{\mathbf{d}}_T\right)$ is the Legendre-Fenchel conjugate of $\mathbb{I}_{C(p_n)}(\mathbf{p}_T)$ of the convex disk C of radius $-\mu p_n$ centered at the origin $C = \{\mathbf{p}_T \mid \|\mathbf{p}_T\| + \mu p_n \leq 0, p_n \leq 0\}$. Extending these laws to (discrete) global formulas as previously, the pure friction laws are:

$$\mathcal{F}_T \in \partial \mathbb{I}^*_{C^p}\left[\mathcal{D}_T\left(\delta\mathcal{U}\right)\right] \qquad \mathcal{D}_T\left(\delta\mathcal{U}\right) \in \partial \mathbb{I}_{C^p}\left[\mathcal{F}_T\right] \tag{44}$$

where $\mathbb{I}^*_{C^p}\left[\mathcal{D}_T\left(\delta\mathcal{U}\right)\right]$ is the conjugate of $\mathbb{I}_{C^p}\left[\mathcal{D}_T\left(\delta\mathcal{U}\right)\right]$.

3. **Unilateral contact with friction.** In unilateral frictional contact, it is known that the normal pressure is coupled with tangential friction. The combination of normal contact and pure friction gives:

$$p_n \in \partial \mathbb{I}_{R_+}\left(d_n\right) \qquad \mathbf{p}_T \in \partial \mathbb{I}^*_{C(p_n)}\left(\dot{\mathbf{d}}_T\right) \tag{45}$$

where the domain of friction criterion C depends on p_n. The extension of local laws to (discrete) global formulas as previously gives:

$$\mathcal{F}_n \in \partial \mathbb{I}_{\mathbb{R}^p_+}[\mathcal{D}_n(\mathcal{U})] \qquad \mathcal{F}_T \in \partial \mathbb{I}^*_{C^p}[\mathcal{D}_T(\delta\mathcal{U})] \qquad (46)$$

where each domain C depends on $\mathcal{F}_n$.

4. **Thermocontact with friction.** Heat conduction laws at the interface of two solids were proposed in the past [14]:

$$q' = -\kappa \frac{\delta\theta}{\theta'} \qquad \kappa = \hat{\kappa}(p_n)$$

where κ is a heat conduction coefficient at the interface. In the framework of normal dissipation, we can propose the potential [19]:

$$\psi_c = \frac{1}{2}\hat{\kappa}(p_n)\left(\frac{\delta\theta}{\theta'}\right)^2 \qquad q' = \hat{\kappa}(p_n)\frac{\delta\theta}{\theta'}$$

For a contact thermomechanics law with friction, we combine the above laws to obtain the interfacial energy and potential dissipation [9]:

$$\hat{\phi}_c(d_n) = \mathbb{I}_{R_+}(d_n)$$

$$\hat{\psi}_c\left(\dot{\mathbf{d}}_T, \frac{\delta\theta}{\theta'}; p_n\right) = I^*_{C(p_n)}(\dot{\mathbf{d}}_T) + \frac{1}{2}\hat{\kappa}(p_n)\left(\frac{\delta\theta}{\theta'}\right)^2 \qquad (47)$$

3   Numerical Algorithms

We limit to elastic solids subject to conservative external forces and dry frictional contact. Let ψ be the strain energy of B. As an illustration, the constitutive laws of transversely isotropic bone are defined by [15]:

$$\psi = \frac{1}{2}e_1\text{tr}^2\varepsilon + \frac{1}{2}e_2\text{tr}\varepsilon^2 + e_3\text{tr}\mathbf{M}\varepsilon\text{tr}\varepsilon + e_4\text{tr}\mathbf{M}\varepsilon^2 + \frac{1}{2}e_5\text{tr}^2\mathbf{M}c \qquad (48)$$

$$\sigma = -\rho\frac{\partial\psi}{\partial\varepsilon} \qquad \varepsilon = \frac{1}{2}\left(\text{grad}\mathbf{u} + \text{grad}\mathbf{u}^T\right) \qquad (49)$$

in which $\mathbf{M} = \mathbf{m} \otimes \mathbf{m}$ is the structural tensor ($\mathbf{m}$: transverse isotropy direction) and where the elastic coefficients may be related to elastic constants [16]:

$$e_1 = \left(\nu_t + \nu_l^2 \frac{E_t}{E_l}\right)\frac{E_t}{\Delta} \qquad e_2 = \frac{E_t}{1 + \nu_t}$$

$$e_3 = \left(\nu_l - \nu_t + \nu_l\nu_t - \nu_l^2\frac{E_t}{E_l}\right)\frac{E_t}{\Delta} \qquad e_4 = 2\mu_l - \frac{E_t}{1 + \nu_t}$$

$$e_5 = -4\mu_l + \left(1 - 2\nu_l - 2\nu_l\nu_t - \nu_l^2\frac{E_t}{E_l} + \left(1 - \nu_t^2\right)\frac{E_l}{E_t}\right)\frac{E_t}{\Delta} \qquad (50)$$

where $\Delta = \left(1 - \nu_t - 2\nu_l^2\frac{E_t}{E_l}\right)(1 + \nu_t)$. The same relations hold for B' but, constitutive equations may be simplified if B' is homogeneous and isotropic.

3.1   *Variational formulation*

For deriving the (quasistatic) equations governing thermocontact, we proceed step by step. First, for an elastic solid B, the total energy $\Psi(\mathbf{u})$ is equal to the internal energy minus the external potential energies

$$\Psi(\mathbf{u}) = \frac{1}{2}\int_B E^{ijkl}\varepsilon_{ij}(\mathbf{u})\,\varepsilon_{kl}(\mathbf{u})\,dv - \int_B \rho\mathbf{b}\cdot\mathbf{u}dv - \int_{\partial B_p} \mathbf{p}_n\cdot\mathbf{u}da \qquad (51)$$

The unconstrained minimization problem would be characterized by the necessary condition (equilibrium equation of B) $\nabla_{\mathbf{w}}\Psi(\mathbf{u}) = 0, \forall\mathbf{w}\in W$, W space of kinematically admissible virtual velocity. Assuming now that B is constrained by unilateral contact on some part of its boundary, we assume dicretized for simplifying (Fig. 1). In presence of inequality constraint $\mathcal{D}_n(\mathcal{U})\in\mathbb{R}_+^p$, $\mathbb{R}_+^p = R_+\times...\times R_+$ (p-times), minimization problem can be reformulated into a nonconstrained minimization by adding the indicator function to the total energy:

$$\min_{\mathcal{U}}\left\{\Psi(\mathcal{U}) + \mathbb{I}_{\mathbb{R}_+^p}(\mathcal{D}_n)\right\} \qquad (52)$$

The solution of the nondifferentiable problem can be characterized by calculating the (sub)-derivative to give (necessarily condition of extremality):

$$0 \in \nabla_{\mathcal{W}}\Psi(\mathcal{U}) + [\nabla_{\mathcal{W}}\mathcal{D}_n(\mathcal{U})]^T \,\partial\mathbb{I}_{\mathbb{R}_+^p}(\mathcal{D}_n) \qquad \forall\mathcal{W}\in W \qquad (53)$$

For two elastic solids subject to conservative external loads, $\mathcal{U}$ includes the two displacement fields of B and B'. The total energy $\Psi \equiv \Psi_B + \Psi'_{B'}$. For solids in unilateral contact, the problem solution becomes:

$$\mathcal{U}^* = \arg\min_{\mathcal{U}}\left\{\Psi(\mathcal{U}) + \mathbb{I}_{\mathbb{R}_+^p}[\mathcal{D}_n(\mathcal{U})]\right\} \qquad (54)$$

where $\mathbb{I}_{\mathbb{R}^p_+}[x]$ is the indicator function of $\mathbb{R}^p_+ = R_+ \times \ldots \times R_+$. For solids in contact with Coulomb pure friction, the formulation is incremental due to the path dependence,

$$\delta \mathcal{U}^* = \arg \min_{\delta \mathcal{U}^*} \{ \Psi(\mathcal{U} + \delta \mathcal{U}) + \mathbb{I}^*_{C^p}[\mathcal{D}_T(\delta \mathcal{U})] \} \qquad (55)$$

where $\mathbb{I}^*_C[x]$ is the Fenchel conjugate of $\mathbb{I}_C[x]$ and the set C^p the Cartesian product of p local disks C, with fixed radius. For solids in unilateral contact with Coulomb friction, we have the coupling effects

$$\delta \mathcal{U}^* = \arg \min_{\mathcal{U}} \{ \Psi(\mathcal{U} + \delta \mathcal{U}) + \mathbb{I}_{\mathbb{R}^p_+}[\mathcal{D}_n(\mathcal{U} + \delta \mathcal{U})] + \mathbb{I}_{C^p(\mathcal{U}^*)}[\mathcal{D}_T(\delta \mathcal{U})] \} \quad (56)$$

where $C^p(\mathcal{U})$ is the Cartesian product of local disks $C^p(p^*_n)$ whose radius $-p^*_n$ depend on the solution $\mathcal{U}^*$. Therefore, the frictional contact problem is called quasi-optimization problem because $C^p(\mathcal{U}^*)$ depends on solution [8].

3.2   Methods for solving contact mechanics

Two classes of methods are adopted for solving the interface problem.

1. For *penalty method*, $\mathbb{I}_{\mathbb{R}^p_+}[\mathcal{D}_n(\mathcal{U})]$ is approximated by the quadratic function $\frac{r}{2} D^2_{\mathbb{R}^p_+}[\mathcal{D}_n(\mathcal{U})]$, where $D_C(x)$ is the distance between x and C. r is the penalty parameter. This gives the penalized potential to be minimized for unilateral contact:

$$\Pi(\mathcal{U}) \equiv \Psi(\mathcal{U}) + \frac{r}{2} D^2_{\mathbb{R}^p_+}[\mathcal{D}_n(\mathcal{U})] \qquad (57)$$

    Solutions of contact carried systematic errors inherent to the method itself [8] (depending on r). In biomechanics where the calculation of variables (micromotions, stresses, and temperature) requires a high precision, it is deemed necessary to develop most accurate computational methods. Another basic motivation for developing Lagrange multiplier methods is to avoid the ill-conditioning associated with the usual penalty methods. r is not a physical parameter.

2. Multiplier methods start by introducing dual variables $\mathcal{F}_n$ (same dimension as $\mathcal{D}_n$). At the equilibrium, dual variables equal the contact reaction at the interface. For *augmented Lagrangian method*, the first step is to define the standard Lagrangian functional:

$$\Pi(\mathcal{U}) \equiv \Psi(\mathcal{U}) + \mathcal{F}_n \cdot \mathcal{D}_n(\mathcal{U}). \qquad (58)$$

    An intermediate step is to replace the problem of minimization of $\Psi(\mathcal{U})$ under the nonlinear inequality $\mathcal{D}_n(\mathcal{U}) \in \mathbb{R}^p_+$ by the a mixed $min - max$

problem under linear dual inequality:

$$\min_{\mathcal{U}} \max_{\mathcal{F}_n \leq 0} \{\Psi(\mathcal{U}) + \mathcal{F}_n \cdot \mathcal{D}_n(\mathcal{U})\} \tag{59}$$

To show this, recall that the indicator function $\mathbb{I}_{\mathbb{R}_+^p}(\mathcal{D}_n)$ is the Legendre-Fenchel conjugate of $\mathbb{I}_{\mathbb{R}_-^p}(\mathcal{F}_n)$ [13]:

$$\mathbb{I}_{\mathbb{R}_+^p}(\mathcal{D}_n) = \sup_{\mathcal{F}_n} \left\{\mathcal{D}_n \cdot \mathcal{F}_n - \mathbb{I}_{\mathbb{R}_-^p}(\mathcal{F}_n)\right\} \tag{60}$$

In the special case where we deal with indicator function [13], the expression "$\mathcal{D}_n \cdot \mathcal{F}_n - \mathbb{I}_{\mathbb{R}_-^p}(\mathcal{F}_n)$" takes the value $-\infty$ when $\mathcal{F}_n \notin \mathbb{R}_-^p$, then we have the relation:

$$\sup_{\mathcal{F}_n} \left\{\mathcal{D}_n \cdot \mathcal{F}_n - \mathbb{I}_{\mathbb{R}_-^p}(\mathcal{F}_n)\right\} = \sup_{\mathcal{F}_n \in \mathbb{R}_-^p} \{\mathcal{D}_n \cdot \mathcal{F}_n\} \tag{61}$$

If we work with $\mathbf{R} \cup \infty$ rather than $\mathbf{R}$ the supremum could be replaced by maximum. Therefore, the minimization problem with inequality constraint can be converted into a standard Lagrangian saddle point one:

$$\min_{\mathcal{U}} \max_{\mathcal{F}_n \in \mathbb{R}_-^p} \{\Psi(\mathcal{U}) + \mathcal{F}_n \cdot \mathcal{D}_n(\mathcal{U})\} \tag{62}$$

Both $\mathcal{D}_n$ and $\mathcal{F}_n$ are unknown variables. The third step consists in relaxing the constraint in the Lagrangian multiplier $\mathcal{F}_n$ by adding a penalization. This gives the augmented Lagrangian functional [8,17]:

$$\Pi_r(\mathcal{U}, \mathcal{F}_n) \equiv \Psi(\mathcal{U}) - \frac{1}{2r}\|\mathcal{F}_n\|^2 + \frac{1}{2r}D^2_{\mathbb{R}_+^p}[\mathcal{S}_n(\mathcal{U})] \tag{63}$$

in which r is a penalty parameter, $\mathcal{S}_n(\mathcal{U}) \equiv \mathcal{F}_n + r\mathcal{D}_n(\mathcal{U})$ the augmented multiplier, and $D_{\mathbb{R}_+^p}[\mathcal{S}_n]$ the distance between $\mathcal{S}_n$ and $\mathbb{R}_+^p$. The *min−max* problem with augmented Lagrangian is not constrained:

$$\min_{\mathcal{U}} \max_{\mathcal{F}_n} \{\Pi_r(\mathcal{U}, \mathcal{F}_n)\} \tag{64}$$

$\Pi_r(\mathcal{U}, \mathcal{F}_n)$ is differentiable with respect to $\mathcal{U}$ and $\mathcal{F}_n$.

3. For unilateral frictional contact, we have the combination of global and incremental formulation due to friction, the augmented Lagrangian functional to be minimized is called quasi-functional since solutions $\mathcal{U}^*$ and $\mathcal{F}_n^*$ appear as arguments through $\mathcal{S}_n^* = \mathcal{F}_n^* + r\mathcal{D}_n^*(\mathcal{U}^*)$:

$$\begin{aligned} \Pi_r(\mathcal{U}, \delta\mathcal{U}, \mathcal{F}_n, \mathcal{F}_T; \mathcal{U}^*, \mathcal{F}_n^*) &= \Psi(\mathcal{U}) + \mathcal{F}_n \cdot \mathcal{D}_n(\mathcal{U} + \mathcal{F}_T \cdot \mathcal{D}_T(\delta\mathcal{U})) \\ &+ \frac{r}{2}\|\mathcal{D}_n(\mathcal{U})\|^2 - \frac{1}{2r}D^2_{R_-^p}[\mathcal{S}_n(\mathcal{U}, \mathcal{F}_n)] \\ &+ \frac{r}{2}\|\mathcal{D}_T(\delta\mathcal{U})\|^2 \\ &- \frac{1}{2r}D^2_{C_-^p(\mathcal{S}_n(\mathcal{U}^*, \mathcal{F}_n^*))}[\mathcal{S}_T(\delta\mathbf{U}, \mathcal{F}_T)] \end{aligned} \tag{65}$$

where $\mathcal{S}_T \equiv \mathcal{F}_T + r\mathcal{D}_T$, and $C^p \equiv C^p\{\mathbb{P}_{R_-^p} [S_n(\mathcal{U}, \mathcal{F}_n)]\}$

3.3   Methods for solving contact thermomechanics

The variational formulation of contact mechanics is extended to include thermal effects. We have to consider the global displacement vector $\mathcal{U}$ and gap vector $\mathcal{D}$, the global temperature vector $\Theta \equiv (\theta, \theta')$ and temperature gap vector $\mathcal{T} \equiv \theta_0' - \theta_0 + \Delta\theta' - \Delta\theta$. As for frictional contact, we face a quasi-optimization problem.

1. For penalty method, the quasi-functional is proposed as:

$$\Pi(\mathcal{U}, \Theta, \mathcal{U}^*, \Theta^*) = \Psi(\mathcal{U}, \Theta, \mathcal{U}^*, \Theta^*)$$
$$+ \frac{r}{2}D_{R_+^p}^2 [\mathcal{D}_n(\mathcal{U})] + \frac{r}{2}h\mathbb{P}_{R_-^p} [\mathcal{D}_n(\mathcal{U}^*)] \ [\mathcal{T}(\Theta)]^2 \ (66)$$

We have a lack of symmetry because heating of an elastic body induces significative dilatation but the converse is not true since an elastic deformation does not increase its temperature significantly. Heat conduction through the contact may be proportional to the contact pressure whereas modification of temperature gap does not influence pressure.

2. For augmented Lagrangian method, the quasi-functional includes the dual variable to give for the unilateral contact model:

$$\Pi(\mathcal{U}, \Theta, \mathcal{U}^*, \Theta^*) = \Psi(\mathcal{U}, \Theta, \mathcal{U}^*, \Theta^*) - \frac{1}{2r}\|\mathcal{F}_n\|^2$$
$$+ \frac{1}{2r}D_{R_+^p}^2 [S_n(\mathcal{U})] + \frac{1}{2}h\mathbb{P}_{R_-^p} [S_n(\mathcal{U}^*)] \mathcal{T}^2(\Theta) \ (67)$$

where $S_n(\mathcal{U}) = \mathcal{F}_n + r\mathcal{D}_n(\mathcal{U})$.

3. For frictional unilateral contact, the augmented Lagrangian quasi-functional holds [18]:

$$\Pi_r(\mathcal{U}, \delta\mathcal{U}, \Theta, \mathcal{U}^*, \delta\mathcal{U}^*, \Theta^*, \mathcal{S}_n^*, \mathcal{S}_T^*) = \Psi(\mathcal{U}, \Theta, \mathcal{U}^*, \Theta^*)$$
$$+ \mathcal{F}_n \cdot \mathcal{D}_n(\mathcal{U}) + \mathcal{F}_T \cdot \mathcal{D}_T(\delta\mathcal{U})$$
$$+ \frac{r}{2}|\mathcal{D}_n(\mathcal{U})|^2 - \frac{1}{2r}D_{R_-^p}^2 [S_n(\mathcal{U})]$$
$$+ \frac{r}{2}\|\mathcal{D}_T(\delta\mathcal{U})\|^2 - \frac{1}{2r}D_{C^p(S_n^*)}^2 [\mathcal{S}_T(\delta\mathcal{U})]$$
$$+ \frac{1}{2}h\mathbb{P}_{R_-^p} [S_n(\mathcal{U}^*)] \mathcal{T}^2(\Theta)$$
$$+ \left\{\mathbb{P}_{C^p[S_T^*]} [\mathcal{S}_T^*] \cdot \mathcal{D}_T^*(\delta\mathcal{U}^*)\right\} \mathcal{T}(\Theta) \ (68)$$

For this latter, the min-max problem of the augmented functional is written as follows:

$$\min_{\mathcal{U},\theta} \max_{\mathcal{F}_n,\mathcal{F}_T} \{\Pi_r \left(\mathcal{U}, \delta\mathcal{U}, \Theta, \mathcal{U}^*, \delta\mathcal{U}^*, \Theta^*, \mathcal{S}_n^*, \mathcal{S}_T^*\right)\} \qquad (69)$$

Details of the variational formulations and the generalized jacobian resulting from these $min - max$ problems may be found elsewhere [18].

4   Thermomechanics of the Cemented Hip Stem

4.1   Clinical problems of cemented hip arthroplasty

After cemented total hip replacement, two types of interface failures occur: failure of stem-cement interface as a result of debonding of stem from cement and failure of bone-cement interface as a result of the gradually failure of bone-cement inter-digitation [2,1,19]. Interface failures could have resulted from the occurrence of abnormally high shear and compressive stresses within the cement and excessive debonding and slipping at both stem cement and bone-cement interfaces. At the bone-cement interface, penetration of the cement into trabecular bone leads to a complex bone-cement interface which also has a tensile and shear interfacial strengths. Despite their geometrical and mechanical complexity, we can assume a unilateral frictional contact for both the bone-cement and cement-prosthesis interfaces. Many factors have been implicated in the distribution of stress and micromotion and among others, friction coefficients at both stem-cement and bone-cement interfaces [20,2], stem stiffness, cement thickness [3,21]. Hence, the goal of this section were to investigate, with the previous models, the effects of the roughness between the bone and the cement, the stem stiffness, and the cement thickness on the cement stress and on the interfacial micromotions after cemented THR. We also investigate the effects of cement thickness on the temperature distribution.

4.2   Finite element model of the cemented hip stem

A collarless, straight, symmetric hip stem with a cement mantle of uniform thickness was digitized and numerically inserted in the reconstructed femur following typical surgical procedure [22] (see Fig. 2). We considered a titanium stem $(Ti6Al4V)$ (Young's modulus $E = 110000MPa$, Poisson's ratio: $\nu = 0.3$) and a chromium-cobalt stem (Young's modulus: $E = 200000MPa$, Poisson's ratio: $\nu = 0.3$). For each of them, two interfaces (stem-cement and bone-cement) were modelled as unilateral frictional thermocontact. The Young's modulus of the cement was $2200MPa$. The loading conditions corresponded to the single limb stance situation of a gait cycle. The load bearing

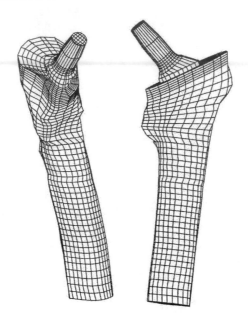

Figure 2. Finite element mesh of the bone-implant system (Anterior-posterior view). The 3D geometry of the femur was reconstructed via Quantitative Computed Tomography.

on the femoral head was simulated with a force of magnitude three times body weight (Patient weight: $600N$) and decomposed into axial, in-plane and out-plane directions [22]. Muscle forces (gluteus minimus, medius, maximus and psoas) were implemented in the model [23,22].

1. The *mechanical analysis* includes three points. First the titanium alloy implant was compared to the chromium-cobalt implant. The cement thickness was set to $4mm$ in this case. The friction coefficients at the bone-cement interface and at the stem-cement interface were set to $\mu_{bc} = 1.0$ and $\mu_{sc} = 0.40$ respectively. Secondly, the sensitivity of the results with respect to the cement thickness was investigated. The cement thicknesses of $2mm$; $3mm$; $4mm$; $5mm$; $7mm$ were used. For regions where the bone thickness did not allow having the prescribed cement thickness, the maximum distance between the implant outer surface and the cortical endosteal femur was used instead. Thirdly, the effects of bone-cement interface rugosity were investigated. To this end, the friction coefficient at the bone-cement interface was set successively: 0.4,

0.6, 0.8 and 1.0. For this example, the thickness of the cement mantle was set $4mm$ [22].

2. For *thermal analysis*, we chose the chromium-cobalt implant inserted with cement of uniform thickness. Two interfaces (stem-cement: $\mu_{sc} = 1.0$; cement-bone: μ_{sc} variable) were considered. The distribution of temperature within the cemented bone-implant system was simulated with different cement thickness: $2mm$, $3mm$, $4mm$ and $5mm$. For the initial and boundary conditions, the temperature of the implant and the bone is initially set as $37°$. The temperature of the orthopaedic cement was held uniform and constant at $100°$. The outer surface of the femur was kept at the body temperature $37°$.

4.3 Micromotions and stresses at the interfaces

1. Titanium alloy versus Cobalt Chromium alloy

   Titanium implant. Distribution of micromotions (debonding and slipping) at the bone-cement interface were calculated. Slipping was higher than $30\mu m$ at the proximal lateral, intermediate medial and distal lateral regions of the bone-cement interface. The peak slipping value was around $67\mu m$. The debonding was in general less than $10\mu m$ over almost the entire region of the interface. Nevertheless, the magnitude could exceed $30\mu m$ in the proximal medial and distal lateral regions (peak: $35\mu m$). At stem-cement interface, slipping exceeded $30\mu m$ over most of the proximal region and then decreased gradually towards the distal part. Concerning the stresses at the bone-cement interface, shear stress exceeded $1.0MPa$ at the proximal lateral, intermediate medial and distal medial regions (peak: $2.4MPa$). Elsewhere, the magnitude remained lower than $0.5MPa$. High compressive stress (peak: $4.4MPa$) occurred in the same locations as the high shear stress (peak: $4.4MPa$). At the stem-cement interface, the peak shear stress was $3.0MPa$ and the peak compressive stress was approximately $7.0MPa$.

   Cobalt chromium implant. At the bone-cement interface, high slipping occurred at the proximal lateral, intermediate medial and distal lateral regions of the interface (Peak values: $68\mu m$). Debonding magnitude was less than $30\mu m$ over the entire interface (Peak value: $28\mu m$). Peak values of micromotions and stresses at stem-cement interface are reported on Table 2 (Peaks debonding : $63\mu m$, slipping: $109\mu m$, compressive stress: $6.6MPa$, shear stress: $2.6MPa$). Peak values of shear and compressive

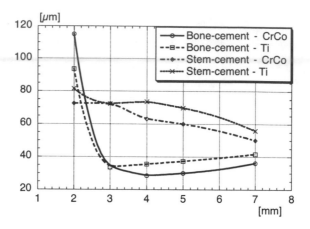

Figure 3. Maximum values of normal relative micromotions $\max_{\partial B_c}\{\mathcal{D}_n\}$ at the bone-cement and cement-stem interfaces related to the thickness of the cement.

stress were higher than $1MPa$ and had nearly the same distribution as for the titanium stem (Peak stress: compressive: $4.2MPa$; shear: $2.5MPa$).

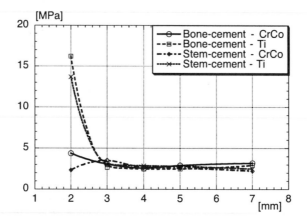

Figure 4. Maximum values of interfacial shear stresses at the bone-cement and cement-stem interfaces related to the thickness of the cement mantle.

2. Influence of the cement thickness

For both implants, debonding at bone-cement interface increased significantly for a cement $2mm$ thick (cobalt-chromium: $115\mu m$, titanium: $93\mu m$). For a thickness higher than $3mm$, debonding was less than $30\mu m$ over almost the entire interface. At stem-cement interface, maximal debonding decreased slightly with the cement thickness. It was observed that for titanium implant a minimal of debonding peak values ($33\mu m$) occurred when the cement thickness was approximately $3mm$ (Fig. 3) and for chromium cobalt stem, minimum debonding ($28\mu m$) occurred for a cement thickness of $4mm$ (Fig. 3). The cobalt-chromium implant in general had lower debonding than the titanium implant.

Evolution of slipping at the bone-cement and stem-cement interfaces vs. cement thickness are reported in figure 5. The implant behaviors (titanium and chromium-cobalt) were quite similar. For a cement thickness of less than $3mm$, slipping increased drastically and exceeded $100\mu m$ over the entire bone-cement interface (Titanium: $1650\mu m$, cobalt-chromium $680\mu m$). The peak values of slipping were minimalized when the thickness was $3mm$ ($52\mu m$ for both implants). For a thickness in the range of $3mm, 7mm$, slipping remained constant with highest values occurring in the proximal and distal lateral, and in the medial regions of the bone-cement interface. For thicknesses greater than $7mm$, the peak values of shear micromotions increased to $170\mu m$. In all cases, slipping at the bone-cement interface was greater than slipping at the stem-cement interface.

Regarding the stress, peak values of shear and compressive stresses at both interfaces (bone-cement and stem-cement) for thickness greater than $3mm$ were not significantly influenced by cement thickness in either implants. For $2mm$ thick layer of cement, abnormally high compressive (Fig. 5) and shear stresses (Fig. 4) were observed at both interfaces for the titanium stem while only a slight increase in stress was noticed for the cobalt-chromium stem. Overall, the shear stress was lower than the compressive stress.

3. Influence of the friction coefficient

The magnitude of slipping for the friction coefficient at the bone-cement interface was higher than $30\mu m$ for the entire interface for the case of a small friction coefficient ($\mu = 0.4$). The maximal values of slipping were higher for a low friction coefficient and remained nearly constant for higher friction coefficients. Conversely, the debonding was minimal for lower coefficient and maximal for higher coefficient. When the coefficient

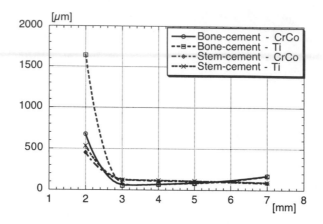

Figure 5. Maximum values of sliding relative micromotions $\max_{\partial B_c}\{\mathcal{D}_T\}$ at the bone-cement and cement-stem interfaces related to the thickness of the cement mantle.

friction of bone-cement interface increased, slipping at the stem-cement interface increased and debonding one decreased (Fig. 6).

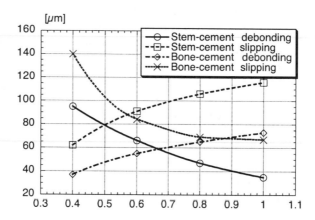

Figure 6. Maximum values of relative micromotions (debonding, sliding) at the bone-cement and cement-stem interfaces related to the friction coefficient.

The compressive stress was minimal for a low friction coefficient ($\mu = 0.4$) and increased gradually with the coefficient while the shear stress remained nearly constant (Fig. 7). Maximal compressive and shear stresses in the cement did not depend significantly on friction coefficient at the bone-cement interface. Bone-cement stresses (compressive and shear) were lower than stem-cement stresses and the shear stress was lower than the compressive stress.

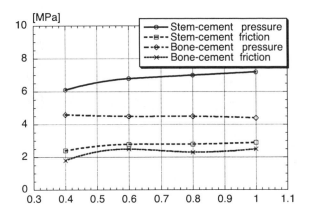

Figure 7. Maximum values of interfacial stresses (pressure, friction) at the bone-cement and cement-stem interfaces related to the friction coefficient.

4.4   Finite element simulation of thermal effects

The distribution of temperature and stress were calculated inside bone, cement, and stem, and at the interfaces (bone-cement and stem-cement) during implantation. Bone, cement and stem properties were taken from [18,22]. Inside structures, peak of temperature was obviously located in the cement. This peak was minimal (76°) when the cement thickness was the lowest $2mm$, and maximal (90°) for a cement mantle of $5mm$ (Fig. 8). At the bone-cement interface, the calculation has shown that temperature increases also with thickness. Stress and temperature were lowest (50°) for a cement mantle thickness of $2mm - 3mm$ and maximal (63°) for a thickness of $5mm$. Moreover, a greater part of the heat generated by polymerization was absorbed by cement and stem. It is attempting to relate the evolution of bone-cement temperature to the probability of late failure at this interface.

The present thermal study has shown that for a cement thickness of $2mm - 3mm$, the peak of temperature inside the bone was minimal. So, even the heat generated by cement may cause bone necrosis in the surrounding tissue, an optimal cement thickness might reduce the depth of penetration of this damage inside bone. Such a result should be improved by introducing the kinetics of the cement polymerization [24].

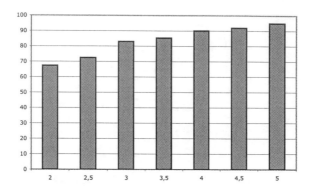

Figure 8. Maximum values of temperature inside the bone-cement-implant system (thickness ranging from 2mm to 5mm).

In addition to temperature field, it was shown that pressure magnitude varies from $+296MPa$ (tensile) and $-97MPa$ (compression). The existence of tensile regions in the vicinity of the interfaces means the occurrence of residual stress when the cement temperature decreases in the course of time. High residual stress near the interfaces is a probable apparition of microcracks density since the mechanical strengths of cement-bone interface, for instance, are respectively $2.25MPa$ and $1.35MPa$ [4].

4.5   Discussion

One should always question if the development of complicated theoretical models and sophisticated numerical algorithms has sense since there are now so many valuable commercial softwares. The following discussion attempts to highlight the usefulness of the previous simulations in the optimization of cement thickness.

1. *Choice of the stem material*

   Comparison of titanium and chromium-cobalt implants (same geometry, friction coefficient, loading condition, and cement thickness) showed that the implant stiffness had no significant effect on micromotion magnitude at the bone-cement interface (Peak: titanium: $67\mu m$, chromium-cobalt:

$68\mu m$). The two stems presented comparable lower peak values the shear stress at bone-cement interface (titanium: $2.4MPa$; chromium-cobalt: $2.5MPa$). High stress occurred over the almost the same area for the two implants. However, a fine look at the results pointed out that a $2mm$ cement thickness drastically increased titanium stem stresses but not for chrome-cobalt, which could lead to cement fracture generating in vivo large particulate debris [25].

2. *Optimization of the cement thickness*

The optimization of the cement thickness during cemented hip arthroplasty remains a challenge for orthopaedic surgeons. Conclusions as to the optimal range of thickness were contradictory. It was shown that a cement mantle thicker than $5mm$ was responsible for the radiolucent lines at the bone-cement interface and a cement mantle lesser than $2mm$ thick induced cement fracture [3]. These findings seemed to be explained numerically by the occurrence of very high shear and compressive stresses [19] which predicted a high risk of cement fracture for a cement less than $3mm$ thick. By accounting for discontinuity at the two interfaces, the present study showed that cement thickness had influence not only on stresses but also on micromotions. A $2mm$ thick layer of cement increased a) shear stress in the proximal part and at the tip of the stem and b) micromotions over the entire bone-cement interface. These outcomes could be related to local bone necrosis and occurrence of osteolysis in these regions due to the presence of the cemented debris resulting from interfacial shear friction [26]. At the other extreme, thicknesses greater than $7mm$ increased slipping at the bone-cement interface and conformed to experimental results [3]. The present study suggested that an optimum cement thickness was in the range of $3mm - 5mm$ where micromotion peaks were minimal (titanium: slipping $52\mu m$, debonding: $33\mu m$; chromium-cobalt: slipping $52\mu m$, debonding $28\mu m$). Shear stress was also lower when thickness was slightly greater than $3mm$.

Recent experimental studies have measured the tensile and shear yield stresses $1.35MPa$ and $2.25MPa$. By the way, previous studies on the interface mechanics admitted "penetration error" of the order $\|\mathcal{D}_n\| \simeq 10\mu m$ by using penalty method [7]. In the worst case (cement thickness $e = 2mm$), double penetration at both bone-cement and cement-stem interface gives an error of $2\|\mathcal{D}_n\| \simeq 20\mu m$, which corresponds to a radial cement strain $\varepsilon_c \simeq 10^3 \frac{\mu m}{m}$. This points out the necessity to accurately calculate the variables at the interface, because the comparison of finite element and experimental strains strongly depends on the penalty parameter. Remembering that bone biology at the bone-cement interface is driven by very small amplitude micromotions [27,28], care should be taken when using non reliable numerical algorithms for solving interfacial thermomechanics.

3. *Choice of the stem surface design*

The choice of surface roughness of the hip stem also remains controversial in orthopaedics. Experimental studies [29] suggested that improving the strength of stem-cement interface should enhance the longevity of the component [30]. However, improved bonding between the cement and the implant might induce early failure of cemented stem [2]. In the same way, numerical simulations showed augmentation of stresses at the bone-cement interface for a fully bonded stem-cement interface compared to partially bonded to the cement [2]. Nonetheless, reduction of stem-cement friction did not decrease shear stresses at the cement-bone interface [19]. Similarly, failure of the bond at the stem-cement interface might initiate the loosening process [31]. This remains a controversial point. By accounting two frictional interfaces, the present example has shown that the decrease of the cement-bone friction coefficient augmented the slipping and decreased the debonding at the bone-cement interface (the stem-cement friction coefficient being maintained constant). In some sense, improvement of the bond at the cement-stem interface significantly increased the relative slipping at the bone-cement interface [22]. Such a phenomenon could be proposed as a biomechanical process promoting early failure at the bone-cement interface [2]. Furthermore, the shear stress magnitude varied slightly with respect to the friction coefficient [19]. The main change was the compressive stress magnitude which was probably due to the hoop effects resulting from the pistoning of the stem in the cement mantle. Such abnormally augmentation of the compressive and hoop stresses certainly induced an overloading of the cement mantle. These results could bring new insight for better understanding biomechanical process of cemented stem failures.

4. *Influence of thermics*

Temperature distribution within the bone-implant system is of central interest for cemented arthroplasty. High temperature may result in thermal injury of surrounding bone. In the same way, the coupling of mechanical and thermal variables is important since the temperature difference at various region arising after orthopaedic cement polymerization induces residual stresses at the interfaces. Namely, the difference between thermal expansion coefficients of the bone, the cement, and the stem are so different and themselves different from the elastic modulus of the stem that stresses are residual concentrated at the interfaces. Experimental measurements have shown significant decreasings of residual stresses on the fracture energies of the bone-cement and cement-stem interfaces [32]. Thermal analysis allows to evaluate temperature distribution during cemented total hip replacement, in order to better understand if the chosen cement mantle thickness presents risks of damage to the tissues surrounding the implant. For thermal aspects, improvements of finite element models by accounting non uniform cement thickness, cement kinetic reaction and other factors [24], should be considered to more accurately evaluate heat generation and temperature distribution. It was also demonstrated that cement cracks resulting from the shrinkage stresses [6] was a most probable source of cement weakness prior to functional loading. Depending on cement temperature and polymerization, stem and bone temperatures, microcrack distribution might be uneven and was denser at the bone-cement interface. In view of all these simplifications, it could be suggested to authors developing finite element interface models to be careful before drawing clinical conclusions whenever interfacial thermomechanics was involved.

Concluding Remarks

Long-term studies of cemented hip arthroplasty have shown that the most frequent cause of failure is the loss of mechanical fixation. Many factors as the bone quality, the implant design and the bone-implant interface are known to influence the quality of fixation of femoral components. Recent investigations have demonstrated that the risk of developing radiolucent lines at the bone-cement interface in the femur was more than 50 percent independently of the stem shape [33]. Evaluation of the bone-implant micromotion and interfacial stress is important for determining the mechanical environment acting on the bone cells [34]. This work is devoted to the developement of thermomechanical models and computational methods to investigate the initial fixation of orthopedic implants. Based on experimental and on clinical studies, [33,27] we assume that radiolucencies at the bone-cement-stem interfaces could be attributed to high shear micromotions. Since the radiolucent line appears for most stem shape designs, a cemented femoral component was chosen to illustrate the developed models by varying different parameters [22]. From biology point of view, radiolucent lines at the bone-cement interface correspond to a fibrous layer, meaning that no close contact forms between the surrounding bone and the cement mantle. This gap is filled with fibroblasts which may be due to bone necrosis but also potentially by cell death by apoptosis [35] or demineralization [34]. Tissue change often appears after long durations but most probably [33]. Mechanical modelling of cement microcracking behaviour [19] and development of biomechanical models simulating the time evolution of this fibrous tissue layer remain a challenge to bioengineers [36] since all of these phenomena may strongly induce implant instability in the long term.

As with most models, the present one had some limitations. For instance, one should analyze the influence of the stem size and shape, namely the presence of sharp edge, for fixation quality. Important surgical parameters such as a varus orientation of the femoral stem might also influence the non-uniformity of the cement thickness. These problems should be addressed for clinical situations. Among them, the loading case corresponding to single limb stance phase did not constitute the most severe case allowing optimization of cemented femoral component. Use of other loading such as as stair climbing or standing from a chair could give different conclusion. Another limitation was the interface model we used at bone-cement interface. Due to the flow of cement within trabecular bone, the tensile strength at this interface should not be neglected and therefore, a more sophisticated model of stress failure index might be used [19]. Another limitation was the assumption that bone has constant mechanical properties during the course of time. Combination of micromotions and stress analysis with bone density adaptation would be a future extension of the present study. For finite element modelling, there is a controversial assumption about the possibility of debonding at the stem-cement interface [5]. We would like to mention that this interface may be partially or totally bonded immediatly after implantation. Therefore the assumption of totally or partially bonding remains a theoretical approach during the course of time. However, these two extreme hypotheses constitute the two limits to be considered. Varying the friction coefficient at the interfaces [22] or simulating the crack propagation at the interface [19] would be complementary approaches for further investigations.

In summary, contact problems are central to biomechanics of bone-implant interfaces because contact is the principal mean for transfering loads from the implant to bone and vice versa. Thermal stresses due to heat generation and/or heat transfer accross the interface may also decrease the reliability of the cemented implant anchorage. This typically occured for cemented hip replacement. The accuracy of calculation at the interfaces is crucial for reliability of numerical methods and results in this domain. This is demanded by the very high precision of the biological environment at the interfaces: sensitivity relative to the range of shear micromotions for fibrous tissue evolution [36], range of stress stimulus for bone remodelling, range of temperature for bone necrosis. Apart from the unilateral boundary conditions, thermomechanical contact problems present difficulties because of the fact that the extent and the location of the contact area changes during loading and heating. Sometimes, it appears that resolution of computational thermocontact should lie on more fundamental numerical algorithms (consistency) rather than on choosing some pre-implemented solvers and pre-implemented contact elements in pre-existing softwares. In most cases, the use of augmented Lagrangian methods in contact thermomechanics at the bone-implant interface is justified since mixed penalty-duality methods are known to be more robust than primal penalty methods [8,37]. Multiplier methods have the great advantages to give exact solutions of the thermocontact models independently on the algorithm parameters. From mathematics of view (finite dimension), existence theorem was established [37] under most special situations for contact problems with Coulomb friction. In situations where large amplitude slidings are combined with thermal frictional contact as in head-cup interface after total hip replacement [38], thermomechanics interface models should include large slip capability [17,39] and reformulation of the equilibrium equations should account for the large amplitude motion [40].

Acknowledgments

Scientific advises of Prof A. Curnier (EPF Lausanne) are mostly appreciated during the development of the augmented Lagrangian methods.

References

1. Jasty, M., Maloney, W. J., Bragdon, C. R., O'Connor, D., Haire, T., Harris, W. H., The initiation of failure in cemented femoral components of hip arthroplasties, *J Bone Joint Surg* [Br] **73-B** (1991) pp. 551–8.
2. Gardiner, R. C., Hozack, W. J., Failure of the cement-bone interface: A consequence of strengthning the cement-prosthesis interface, *J Bone Joint Surg* [Br] **76-B** (1994) pp. 49–4.
3. Ebramzadeh, E., Sarmiento, A., McKellop, H. A., Llinas, A., Gogan, W., The cement mantle in total hip arthroplasty: analysis of long-term radiographic results, *J Bone Joint Surg* [Am] **76-A** (1994) pp. 77–11.
4. Mann, K. A., Werner, F. W., Ayers, D. C., Mechanical strength of the cement-bone interface is greater in shear than in tension, *J Biomechanics* **32** (1999) pp. 1251–1254.

5. Stolk, J., Verdonschot, N., Cristofolini, L., Toni, A., Huiskes, R., Finite element and experimental models of cemented hip joint reconstructions can produce similar bone and cement strains in pre-clinical tests, *J Biomechanics* **35** (2002) pp. 499–510.

6. Orr, J. F., Dunne, N. J., Quinn, J. C. Shrinkage stresses in bone cement, *Biomaterials* **24** (2003) pp. 2933–2940.

7. Viceconti, M., Muccini, R., Bernakiewicz, M., Baleani, M., Cristofolini, L., Large-sliding contact elements accurately predict levels of bone-implant micromotion relevant to osseointegration, *J Biomechanics* **33** (2000) pp. 1611–1618.

8. Alart, P., Curnier, A., A mixed formulation for frictional contact problems prone to Newton like solution methods, *Comp Meth Appl Mech Engng* **93**(3) (1991) pp. 353–375.

9. Ramaniraka, N. A., Rakotomanana, L. R., Models of continuum with microcrack distribution. Math Mech Solids **5** (2000) pp. 301–336.

10. Zmitrowicz, A., A thermodynamical model of contact, friction and wear: III Constitutive equations for friction, wear and frictional heat, *Wear* **114** (1987) pp. 199–201.

11. Rakotomanana, L. R., A geometric approach to thermomechanics of dissipating continua, *Birkhauser*, Boston, (2003).

12. Starmans, F. J. M., Brekelmans, W. A. M., Janssen, J. D., A thermodynamic theory for a restricted class of contact behaviour, *Int J Solids Structures* **26**(11) (1990) pp. 1287–1300.

13. Moreau, J. J., On unilateral constraints, friction, plasticity, In Capriz G, Stampacchia G. eds. New variational techniques in mathematical physics (CIME, II-73, Edizioni Cremonese, Roma) (1974), pp. 175–322.

14. Wriggers, P., Miehe, C., Contact constraints within coupled thermomechanical analysis — A finite element model, *Comp Meth Appl Mech Engng* **94** (1994) pp. 301–319.

15. Rakotomanana, R. L., Curnier, A., leyvraz, P. F., An objective anisotropic elastic plastic model and algorithm applicable to bone mechanics, *Eur J mech A/solids* **10**(3) (1991) pp. 327–342.

16. Rakotomanana, R. L., Leyvraz, P. F., Curnier, A., Heegaard, J. H., Rubin, P. J., A finite element model for evaluation of tibial prosthesis-bone interface in total knee replacement, *J Biomechanics* **25** (1992) pp. 1413–1424.

17. Heegaard, J. H., Curnier, A., An augmented lagrangian method for discrete large-slip contact problems, *Int J Num Methods Eng* **36** (1993) pp. 569–593.

18. Ramaniraka, A. N., Thermomécanique des contacts entre deux solides déformables. PhD thesis disseration, Dept. of Mech Engng, EPF Lausanne, (1997).

19. Verdonschot, N., Huiskes, R., Cement debonding process of total hip arthroplasty stems, *Clin Orthop* **336** (1997) pp. 297–11.

20. Harrigan, T. P., Harris, W. H., A three-dimensional non linear finite element study of the effects of cement-prosthesis debonding in cemented femoral total hip components, *J Biomechanics* **24** (1991) pp. 1047-12.

21. Huiskes, R., The various stress patterns of press-fit, ingrown and cemented femoral stems, *Clin Orthop* **261** (1990) pp. 27–12.

22. Ramaniraka, N. A., Rakotomanana, L. R., Leyvraz, P. F., The fixation of the cemented femoral component: Effects of stem stiffness, cement thickness and roughness of the cement-bone surface, *J Bone Joint Surgery* **82-B**(2) (2000) pp. 297–303.

23. Crownshield, R. D., Brand, R. A., A physiologically based criterion of muscle force prediction in locomotion, *J Biomechanics* **11** (1981) pp. 793–9.

24. Hansen, E., Modelling heat transfer in a bone-cement-prosthesis system, *J Biomechanics* **36** (2003) pp. 787–795.

25. McKellop, H. A., Sarmiento, A., Schwinn, C. P., Ebramzadeh, E., In vivo wear of titanium-alloy hip prostheses, *J Bone Joint Surg* [Am] **72-A** (1990) pp. 512-6.

26. Willert, H. G., Bertram, H., Buchhorn, G. H., Osteolysis in alloarthroplasty of the hip: The role of bone cement fragmentation, *Clin Orthop* **258** (1990) pp. 108–14.

27. Farron, A., Rakotomanana, R. L., Zambelli, P. Y., Leyvraz, P. F., Total knee prosthesis. Clinical and numerical study of micromovements of the tibial implant, *Rev Chir Orthop Reparatrice Appar Mot* **80** (1995) pp. 28–35.

28. Jasty, M., Bragdon, C., Burke, D., O'Connor, D., Lowenstein, J., Harris, W. H., In vivo skeletal responses to porous-surfaced implants subjected to small induced motions, *J Bone Joint Surg* [Am] **79** (1997) pp. 707–714.

29. Bundy, K. J., Penn, R. W., The effect of surface preparation on metal/bone cement interfacial strength, *J Biomed Mater Res* **21** (1987) pp. 773–33.

30. Harris, W. H., Is it advantageous to strengthen the cement-metal interface and use a collar for cemented femoral components of total hip replacement? *Clin Orthop* **285** (1992) pp. 67–6.

31. Mohler, C. G., Callaghan, J. J., Collis, D. K., Johnston, R. C., Early loosening of the femoral component at the cement-prosthesis interface after total hip replacement, *J Bone Joint Surg* [Am] **77-A** (1995) pp. 1315–8.

32. Zor, M., Kucuk, M., Aksoy, S., Residual stress effects on fracture energies of cement-bone and cement-implant interfaces, *Biomaterials* **23** (2002) pp. 1595–1601.
33. Ebramzadeh, E., Normand, P. L., Sangiorgio, S. N., Llinás, A., Gruen, T., McKellop, H. A., Sarmiento, A., Long-term radiographic changes in cemented hip arthroplasty with six designs of femoral components, *Biomaterials* **24** (2003) pp. 3351–3363.
34. Pioletti, D. P., Mueller, J., Rakotomanana, L. R., Corbeil, J., Wild, E., Effect of micro-mechanical stimulations on osteoblasts: Development of a device simulating the mechanical situation at the bone-implant interface, *J Biomechanics* **36** (2003) pp. 131–135.
35. Ciapetti, G., Granchi, D., Savarino, L., Cenni, E., Magrini, E., Baldini, N., Giunti, A., In vitro testing of the potential for orthopedic bone cements to cause apoptosis of osteoblast-like cells, *Biomaterials* **23** (2002) pp. 617–627.
36. Buechler, Ph., Pioletti, D. P., Rakotomanana, L. R., Biphasic constitutive laws for biological interface evolution, *Biomech Model Mechanobiology* **1** (2003) pp. 239–249.
37. Pang, J. S., Stewart, D. E., A unified approach to discrete frictional contact problems, *Int J Eng Science* **37** (2003) pp. 1747–1768.
38. Bergmann, G., Graichen, F., Rohlmann, A., Verdonschot, N., van Lenthe, G. H., Frictional heating of total hip implants. Part 2. Finite element study, *J Biomechanics* **34** (2001) pp. 429–435.
39. Pietrzak, G., Curnier, A., Large deformation frictional contact mechanics: Continuum formulation and augmented Lagrangian treatment, *Comput meth Appl Mech Engng* **177** (1999) pp. 351–381.
40. Buechler, P., Rakotomanana, L., Farron, A., Virtual power based algorithm for decoupling large motions from infinitesimal strains: Application to shoulder joint biomechanics. *Comp Meth Biomech Biomed Eng* **5**(6) (2002) pp. 387–396.

ANALYSIS OF TRANSIENT BLOOD FLOW PASSING THROUGH MECHANICAL HEART VALVES BY LATTICE BOLTZMANN METHODS

ORLANDO PELLICCIONI AND MIGUEL CERROLAZA

Bioengineering Center (CeBio UCV)
Central University of Venezuela, 50.381 Caracas – Venezuela.
E-mail: orlandoCEBIO@cantv.net

MAURO HERRERA

Vargas Hospital of Caracas. Caracas – Venezuela.

In the present work a computational method for the simulation of sanguineous flow based on the Lattice-Boltzmann Method is detailed (LBM). A full bidirectional fluid-structure coupling including opening and closing of the valves effecting the hemodynamical properties of the fluid by transient displacements of the solid-fluid boundaries is considered. In this chapter theoretical bases are considered and described briefly. Results from two and three dimensional analyses are shown for different commercial mechanical heart valves prostheses.

1 Introduction

The dynamic flow behavior of human heart valves has intrigued many investigators including Leonardo da Vinci [1], who described a technique for making a model of the aortic valve. Since Hufnagel's successful implantation of a prosthetic heart valve in a human in 1952 [2], numerous *in vitro* investigations have been conducted to clarify the flow characteristics of artificial heart valves.

Either due to congenital defects, aging, diseased states or to trauma, several components of the cardiovascular system can become abnormal. Typical abnormalities include malfunctioning heart valves, stenosed or dilated arterial segments, loss of normal cardiac contractility or cardiac rhythm disturbances. In such situations, the physician desires to take corrective action with medical therapy or percutaneous mechanical procedures such as angioplasty or valvuloplasty. However, if such corrective actions are not possible, an alternative is to replace the malfunctioning parts with artificial devices. Implanting cardiac pacemakers to maintain cardiac rhythm is one of the most common examples of implanted prosthetic devices.

The favorable results obtained using Heart valve prostheses in humans continue to stimulate not only improvement of existing artificial valves but also development of designs and materials. Experimental and clinical success with prostheses of various types implies that an artificial valve does not need to resemble its biological predecessor.

The development of flow instabilities due to high Reynolds number flow in artificial heart valve geometries inducing high strain rates and stress often leads to hemolysis and related highly undesired effects. Geometric and functional optimization of artificial heart valves is therefore mandatory. In addition to experimental work in this field it is meanwhile possible to obtain increasing insight into flow dynamics by computer simulation of refined model problems. In this chapter, we report the results of the simulation of transient physiological flows in fixed and moveable geometries similar to some commercial mechanical heart valves. The visualization of emerging complicated flow patterns gives detailed information about the transient history of the systems dynamical stability. Stress analysis indicates temporal shear stress peaks even far away from walls. The mathematical approach used was the Lattice Boltzmann method. Reasonable results for velocity and shear stress fields were obtained in this work. Finally, problems, shortcomings and possible extensions of this approach are discussed.

2   Theory

2.1   Natural heart valves

There are four heart valves which assure that the blood flows in one direction and which play a vital role in maintaining the normal cardiac output and perfusion pressures throughout the cardiac system. Valvular heart diseases predominantly occur in the left part of the heart. Hence it is interesting to consider the flow dynamics past the aortic and mitral valves before we consider flow dynamics past valvular prostheses.

The natural heart valves function efficiently with negligible energy loss from blood flow through the valves. However, diseases like rheumatic valvular disease can affect the performance of the valves. Aging can also cause valvular abnormalities. The valves can become stenotic (stiffening of the leaflets) requiring higher proximal pressures before the valves open. Even in fully open position, the valve orifice area may be small, resulting in high pressure loss through the valves. The valves can also become incompetent, resulting in increased back flow (aortic or mitral insufficiency).

2.2   Types of prosthetic valves

Open heart surgery for valve implantations became possible with the introduction of the cardio-pulmonary bypass pump in the middle 1950s. Freshly explanted natural aortic valves from human cadavers, known as homografts or allografts have also been successfully used as replacement valves. These are the only valves entirely consisting of biological material and sewn into place. However, human tissue valves

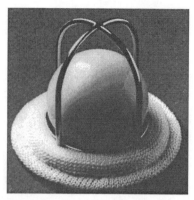

Figure 1. A Starr-Edwards™ caged ball heart valve prosthesis (Courtesy of Edwards Cardiovascular Surgery Division, Baxter Healthcare Corporation, Santa Ana, California).

are difficult to be obtained except in large population areas and therefore it is not a viable option. The first mechanical heart valve was implanted in 1960. Since then, several different valve geometries have been tried. Most valve designs have been empirical, some of them have been successful over the years while others have not become popular. The prosthetic valves currently available today can be broadly classified into two categories: (i) mechanical valves; and (ii) bioprostheses (tissue valve prostheses).

2.2.1   Mechanical Heart valves (MHV)

The currently available mechanical valve prostheses can be classified into three geometries: (i) caged ball, (ii) tilting disc and (iii) bileaflet valve. The Figure 1 shows a caged ball valve. The ball of this valve is made of silicone rubber, and the cage is made of Stellite alloy and has three struts in the aortic valve and four struts in the mitral valve. This valve is centrally occluding and a strong peripheral jet-like flow is observed distal to the valve. Flow separation and vortex formation also occurs downstream from the ball and cage and is present through a major portion of the cardiac cycle.

The Figure 2 shows a Björk-Shiley Monostrut tilting disc valve. The disc is made with pyrolytic carbon (deposited on a graphite substrate). Theses discs have more durability and are chemically inert in the body environment. There are several manufacturers of tilting disc valves with slight variations in the guiding strut and disc shape design. Most of these valves have Stellite orifice rings, guiding struts and Teflon cloth sewing rings. Depending upon the manufacturer, the disc opens to angle from 60° to 75°. The pressure drop measured across a tilting disc valve is smaller than those of the caged ball valve, but slightly larger than that of the bileaflet valve (this   will be discussed   later). The percent regurgitation for the tilting disc valve is larger than that of a caged ball valve but comparable to a bileaflet valve.

Figure 2. A Björk-Shiley Monostrut tilting disck heart valve prosthesis *(Courtesy of Shiley Inc., Irvine, California).*

The Figure 3 shows photographs of a St. Jude Medical bileaflet valve. The valve is entirely made of pyrolytic carbon. In the open position, the bileaflet valves have two peripheral orifices and a central orifice. The pressure drop across the bileaflet valves is the lowest of the mechanical valves and the percent regurgitation is comparable with the tilting disc valves.

2.2.2   Bioprostheses

The biological tissue valves commercially available in the market today are the porcine valve and pericardial tissue valve. Even though the human duramater valve has been tried, it has not become popular. The porcine aortic valve is anatomically similar to that of the human aortic valve. It is harvested, preserved in a gluteraldehyde solution and mounted in supporting struts. A Hancock porcine valve is shown in Figure 4 and it can be observed that the supporting struts protrude into the lumen. The pressure drop across the porcine valve is relatively high and the tissue valve is stenotic especially in smaller size valves. The percentage of regurgitation of the porcine valves is relatively small and is about the same as that of the caged ball valve.

High turbulent shear stresses have been demonstrated to damage the red blood cells and platelets especially in the presence of foreign surfaces. A range of turbulent stresses have been reported in the literature from in vitro measurements in several laboratories. With the mechanical valves, largest turbulent shear stresses have been measured distal to the caged ball valves. The tilting disc valves and the bileaflet valves exhibit lower values of turbulent shear stress. In the tissue valves, the larger turbulent stresses are measured distal to the porcine valve compared to the pericardial valves.

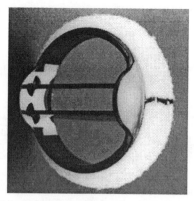

Figure 3. St. Jude Medical bileaflet valve prosthesis; view from the inflow side *(Courtesy of St. Jude Medical Inc., St.Paul, Minnesota).*

3 Development of the lattice Boltzmann equation

Lattice Boltzmann methods (LBM) were developed from Lattice Gas Cellular Automaton (LGCA). To understand the implemented procedure from the Cellular Automaton to the Lattice Boltzmann method and the Boltzmann equation, in this section their theoretical bases are considered and briefly described.

3.1 Cellular Automata Fluids

Cellular automata [3,4] are arrays of discrete cells with discrete values. Yet sufficiently large cellular automata often show seemingly continuous macroscopic behavior [3,5]. Thus they can potentially serve as models for continuum systems, such as fluids. Their underlying discreteness, however, makes them particularly suitable for digital computer simulation and for certain forms of mathematical analysis.

On a microscopic level, physical fluids also consist of discrete particles. But on a large scale, they also seem continuous, and can be described by the partial differential equations of hydrodynamics [6]. The form of these equations is in fact quite insensitive to microscopic details. Changes in molecular interaction laws can affect parameters such as viscosity, but do not alter the basic form of the macroscopic equations. As a result, the overall behavior of fluids can be found without accurately reproducing the details of microscopic molecular dynamics.

A cellular automaton is a discrete dynamic system, which develops by the repeated application of simple deterministic rules. It is defined on a cellular area, among them; one understands an n-dimensional area, which is divided regularly into discrete cells. Each of these cells can accept one of a finite number of z of different conditions. To certain discrete time steps each cell can change its condition as a

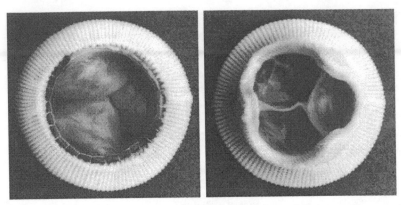

Figure 4.   A Hancock porcine bioprosthetic valve; downstream and upstream view respectively *(Courtesy of Medtronic Heart Valve Division, Irvine, California).*

function of its starting situation and the conditions of its neighboring cells. The rules for these changes are deterministic and uniform in place and time.

3.2   Background of the lattice Boltzmann equation method

Lattice Boltzmann methods are relatively new and quite popular numerical methods for solving the incompressible Navier-Stokes equations of mathematical fluid dynamics in a very special way. Unlike conventional Navier-Stokes solvers, they do not approximate the equations directly but simulate fluid behavior on a mesoscopic level and determine the solution of the incompressible Navier-Stokes equations by computing certain moments of the particle density.

The method of lattice Boltzmann equation (LBE) is an innovative numerical method based on kinetic theory to simulate various hydrodynamic systems [7,8,9]. Although the LBE method was developed only a decade ago, it has attracted significant attention recently, especially in the area of complex fluids including multiphase fluids, suspensions in fluid, and viscoelastic fluids. The lattice Boltzmann equation was introduced to overcome some serious deficiencies of its historic predecessor: the lattice gas automata (LGA). The lattice Boltzmann equation circumvents two major shortcomings of the lattice gas automata: intrinsic noise and limited values of transport coefficients, both due to the Boolean nature of the LGA method. However, despite the notable success of the LBE method in simulating laminar and turbulent flows understanding of some important theoretical aspects of the LBE method, such as the stability of the LBE method, is still lacking. It was only very recently that the formal connections between the lattice Boltzmann equation and the continuous Boltzmann equation and other kinetic schemes were established.

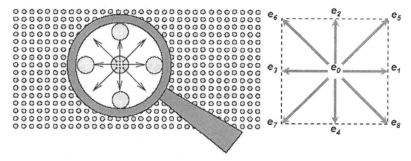

Figure 5. The D2Q9 LBGK model: Each grey circle marks a node of the square lattice. Under the magnifier, we can see the nine groups of particles, marked with black dots, and the corresponding particle velocities *(Figure is taken from Junk et al. [10])*.

3.3 Bhatnagar, Gross, and Krook Model (BGK)

3.3.1 Overview of the lattice BGK equation

In 1986, Frisch *et al.* [11] introduced the lattice Gas method for the incompressible Navier-Stokes equations in two-dimensional space and in the same year d'Humières *et al.* [12] extended the method to three space dimensions. The lattice Boltzmann method was developed in 1988 by McNamara & Zanetti [7] in order to overcome an essential drawback of lattice Gas, namely intrinsic statistical noise, and in 1989, Higuera & Jiménez [13] introduced a linearised version of lattice Boltzmann. In 1992, Chen *et al.* [8] and Qian *et al.* [9] independently proposed to apply the ideas of Bhatnagar *et al.* [14] to the lattice Boltzmann method, thus inventing the lattice BGK (LBGK) method. In 1997, Abe [15] as well as He & Luo [16] independently proved that the governing equations of LBGK can be directly derived from the continuous Boltzmann equation with BGK type collision operator and in 1999, Junk [17] showed the formal connection between the lattice Boltzmann equation and other kinetic schemes. It should also be noted that Luo [18,19] derived a similar set of equations from the continuous Enskog equation and proved consistency of this model to the incompressible Navier-Stokes equations.

3.3.2 Description of the method

We assume that a domain ψ is covered by a regular lattice with equidistant nodes and we denote the set of lattice nodes with χ. At a given time t, we consider the particle density $f(t,x;v)$ of fluid particles located at position and moving with velocity $v \in V_q$, and each one of them contains only a finite number of velocity vectors.

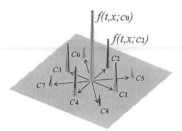

Figure 6. The relative magnitude of particle densities when the fluid is in equilibrium: The particle density corresponding to zero velocity is largest, the common magnitude of particle densities corresponding to velocities in lattice directions is larger than that of particle densities corresponding to velocities in diagonal direction. *(Figure is taken from Junk et al.* [10]*).*

We consider a two-dimensional LBE model with nine discrete velocities (the D2Q9 model [9]) on a square grid with grid spacing δ_x.

In the advection step of the lattice Boltzmann equation, particles move from one node of the grid to one of its neighbors as illustrated in Figure 5.

The discrete velocities are given by

$$
e_\alpha = \begin{cases} (0,0), & \alpha = 0, \\ (\cos[(\alpha-1)\pi/2], \sin[(\alpha-1)\pi/2])c, & \alpha = 1-4, \\ (\cos[(2\alpha-9)\pi/4], \sin[(2\alpha-9)\pi/4])\sqrt{2}c, & \alpha = 5-8. \end{cases} \tag{1}
$$

where $c = \delta_x / \delta_t$, and the duration of the time step δ_t is assumed to be unity. Therefore $c = 1$ in the units of $\delta_x = 1$ and $\delta_t = 1$ (in what follows all quantities are given in non-dimensional units, normalized by the grid spacing δ_x and time step δ_t). Note that non-zero particle velocities are chosen so that particles are moving from a given node to one of its next neighbors in lattice direction or diagonal direction, as illustrated in Figure 5. The Figure 6 visualizes a typical particle density at a given lattice node. At any discrete *time* $t_n = n\,\delta_t$, the LBE fluid is then characterized by the populations of the nine velocities at each node of the computational domain

$$
\left| f(r_j, t_n) \right\rangle \equiv (f_0(r_j, t_n), f_1(r_j, t_n), \dots, f_8(r_j, t_n))^{\mathsf{T}} \tag{2}
$$

where T is the transpose operator. The time evolution of the state of the fluid follows the general equation

$$
\left| f(r_j + e_\alpha \delta_t, t_n + \delta_t) \right\rangle = \left| f(r_j, t_n) \right\rangle + \Omega \left| f(r_j, t_n) \right\rangle \tag{3}
$$

where collisions are symbolically represented by the operator Ω.

An interesting problem with the treatment of the Boltzmann equation is the computation of the collision operator, which makes it under normal conditions a complicated Integro-differential equation. Bhatnagar, Gross and Krook [14] presented therefore a simplified expression for the collision already in 1954. The idea of this simplification is that most of the details from a two-particle interaction does not affect the macroscopic experimental values significantly. If one is not interested in these details, the BGK Approximation can be used as a simplified collision operator:

$$\Omega = -\frac{1}{\tau}\left(f - f^{(0)}\right) \tag{4}$$

τ is a relaxation time which can still be determined nearer; $f^{(0)}$ is the local Maxwell equilibrium distributions, which are dependent on the local speed and the local density at the regarded local point. Under neglect of the outside forces the simplified Boltzmann equation results thus too:

$$\frac{\partial f}{\partial t} + \vec{\xi} \cdot \frac{\partial f}{\partial x} = -\frac{1}{\tau}\left(f - f^{(0)}\right) \tag{5}$$

3.4 The generalized lattice Boltzmann equation (GLBE)

We can make a linear transformation from ψ space to some other space that may be more convenient. In particular, we shall use physically meaningful moments of the quantities f_α that span space **M**. As proposed d'Humières [20], we shall use the generalized lattice Boltzmann equation in which the collision process is executed in moment space **M**.

The mapping between moment space **M** and discrete velocity space ψ is defined by the linear transformation M which maps a vector $|f\rangle$ in ψ to a vector $|m\rangle$ in **M**:

$$|m\rangle = \mathsf{M}|f\rangle \text{ and } |f\rangle = \mathsf{M}^{-1}|m\rangle \tag{6}$$

The advection part of the evolution in Eq. 3 is trivial in ψ, whereas the collision part in Eq. 3 is preferably computed in space **M** both for physical and computational reasons. With reference to kinetic theory of gases [20,21,22] and making use of the symmetries of the discrete velocity set, M is constructed as the following

$$
\mathsf{M} =
\begin{pmatrix}
\langle m_1| \\
\langle m_2| \\
\langle m_3| \\
\langle m_4| \\
\langle m_5| \\
\langle m_6| \\
\langle m_7| \\
\langle m_8| \\
\langle m_9|
\end{pmatrix}
=
\begin{pmatrix}
1 & 1 & 1 & 1 & 1 & 1 & 1 & 1 & 1 \\
-4 & -1 & -1 & -1 & -1 & 2 & 2 & 2 & 2 \\
4 & -2 & -2 & -2 & -2 & 1 & 1 & 1 & 1 \\
0 & 1 & 0 & -1 & 0 & 1 & -1 & -1 & 1 \\
0 & -2 & 0 & 2 & 0 & 1 & -1 & -1 & 1 \\
0 & 0 & 1 & 0 & -1 & 1 & 1 & -1 & -1 \\
0 & 0 & -2 & 0 & 2 & 1 & 1 & -1 & -1 \\
0 & 1 & -1 & 1 & -1 & 0 & 0 & 0 & 0 \\
0 & 0 & 0 & 0 & 0 & 1 & -1 & 1 & -1
\end{pmatrix}
\tag{7}
$$

$$= \left(|m_1\rangle,|m_2\rangle,|m_3\rangle,|m_4\rangle,|m_5\rangle,|m_6\rangle,|m_7\rangle,|m_8\rangle,|m_9\rangle\right)^{\mathsf{T}}.$$

The components of the row vector $\langle m_\beta|$ in matrix M are polynomials of the x and y components of the velocities $\{e_\alpha\}$, $e_{\alpha,x}$ and $e_{\alpha,y}$. The vectors $\langle m_\beta|$, $\beta = 1,2,\ldots,9$, are orthogonalized by the Gram-Schmidt procedure. Specifically,

$$|m_1\rangle_\alpha = \|e_\alpha\|^0 = 1, \qquad\qquad\qquad |m_2\rangle_\alpha = -4\|e_\alpha\|^0 + 3\left(e_{\alpha,x}^2 + e_{\alpha,y}^2\right),$$

$$|m_3\rangle_\alpha = 4\|e_\alpha\|^0 - \frac{21}{2}\left(e_{\alpha,x}^2 + e_{\alpha,y}^2\right) + \frac{9}{2}\left(e_{\alpha,x}^2 + e_{\alpha,y}^2\right)^2, \quad |m_4\rangle_\alpha = e_{\alpha,x}, \qquad\qquad (8)$$

$$|m_5\rangle_\alpha = \left[-5\|e_\alpha\|^0 + 3\left(e_{\alpha,x}^2 + e_{\alpha,y}^2\right)\right]e_{\alpha,x}, \qquad |m_6\rangle_\alpha = e_{\alpha,y},$$

$$|m_7\rangle_\alpha = \left[-5\|e_\alpha\|^0 + 3\left(e_{\alpha,x}^2 + e_{\alpha,y}^2\right)\right]e_{\alpha,y}, \qquad |m_8\rangle_\alpha = \left(e_{\alpha,x}^2 - e_{\alpha,y}^2\right),$$

$$|m_9\rangle_\alpha = \left(e_{\alpha,x} + e_{\alpha,y}\right).$$

The rows of the transformation matrix M are organized in the order of the corresponding tensor, rather than in the order of the corresponding moment. The local state in **M** space:

$$|m\rangle = \left(\rho, e, \varepsilon, j_x, q_x, j_y, q_y, p_{xx}, p_{xy}\right)^{\mathsf{T}} \qquad\qquad (9)$$

can be physically interpreted: $m_1 = \rho$ is the density, $m_2 = e$ is related to the kinetic energy, $m_3 = \varepsilon$ is related to the kinetic energy square. $m_4 = j_x$ and $m_6 = j_y$ are x and y components of the momentum density, $m_5 = q_x$ and $m_7 = q_y$ are proportional to the x and y components of the energy flux, and $m_8 = p_{xx}$ and $m_9 = p_{xy}$ are proportional to the diagonal and off-diagonal components of the viscous stress tensor.

The main difficulty when using the LBE method to simulate a real isotropic fluid is how to systematically eliminate as much as possible the effects due to the symmetry of the underlying lattice. We shall proceed to analyze some simple (but nontrivial) hydrodynamic situations, and to make the flows as independent as possible of the lattice symmetry.

Inspired by the kinetic theory for Maxwell molecules [21], we assume that the non-conserved moments relax linearly towards their equilibrium values that are functions of the conserved quantities. The relaxation equations for the non-conserved moments are prescribed as follows:

$$e^* = e - s_2\left[e - e^{(eq)}\right], \qquad\qquad \varepsilon^* = \varepsilon - s_3\left[\varepsilon - \varepsilon^{(eq)}\right],$$

$$q_x^* = q_x - s_5\left[q_x - q_x^{(eq)}\right], \qquad q_y^* = q_y - s_7\left[q_y - q_y^{(eq)}\right], \qquad (10)$$

$$p_{xx}^* = p_{xx} - s_8\left[p_{xx} - p_{xx}^{(eq)}\right], \qquad p_{xy}^* = p_{xy} - s_9\left[p_{xy} - p_{xy}^{(eq)}\right]$$

where the quantities with and without superscript * are post-collision and pre-collision values, respectively. The equilibrium values of the non-conserved moments in the above equations can be chosen at will, provided that the symmetry of the problem is respected. Are chosen

$$e^{(eq)} = \frac{1}{\langle e|e \rangle}\left[\alpha_2 \langle \rho|\rho\rangle \rho + \gamma_2 \left(\langle j_x|j_x\rangle j_x^2 + \langle j_y|j_y\rangle j_y^2\right)\right]$$

$$= \frac{1}{4}\alpha_2 \rho + \frac{1}{6}\gamma_2\left(j_x^2 + j_y^2\right),$$

$$\varepsilon^{(eq)} = \frac{1}{\langle \varepsilon|\varepsilon \rangle}\left[\alpha_3 \langle \rho|\rho\rangle \rho + \gamma_4\left(\langle j_x|j_x\rangle j_x^2 + \langle j_y|j_y\rangle j_y^2\right)\right]$$

$$= \frac{1}{4}\alpha_3 \rho + \frac{1}{6}\gamma_4\left(j_x^2 + j_y^2\right),$$

$$q_x^{(eq)} = \gamma_1 \frac{\langle j_x|j_x\rangle}{\langle q_x|q_x\rangle}c_i j_x = \frac{1}{2}c_i j_x,$$

$$q_y^{(eq)} = \gamma_1 \frac{\langle j_y|j_y\rangle}{\langle q_y|q_y\rangle}c_i j_y = \frac{1}{2}c_i j_y,$$

$$p_{xx}^{(eq)} = \gamma_1 \frac{1}{\langle p_{xx}|p_{xx}\rangle}\left(\langle j_x|j_x\rangle j_x^2 - \langle j_y|j_y\rangle j_y^2\right) = \frac{1}{2}\gamma_1\left(j_x^2 - j_y^2\right),$$

$$p_{xy}^{(eq)} = \gamma_3 \frac{\sqrt{\langle j_x|j_x\rangle\langle j_y|j_y\rangle}}{\langle p_{xx}|p_{xx}\rangle}\left(j_x j_y\right) = \frac{1}{2}\gamma_3\left(j_x j_y\right).$$

$$\tag{11}$$

The values of the coefficients in the above equilibria (c_1, $\alpha_{2,3}$, and $\gamma_{1,2,3,4}$) which will be summarized in Sec. 3.5.1.

3.5 Multiple-relaxation-time lattice Boltzmann equation

The relaxation lattice Boltzmann equation (RLBE) was introduced by Higuera & Jiménez [13] to overcome some drawbacks of lattice Gas Automata (LGA) such as large statistical noise, limited range of physical parameters, non-Galilean invariance, and implementation difficulty in three dimensions. In the original RLBE the equilibrium distribution functions and the relaxation matrix were derived from the underlying LGA models. It was soon realized that the connection to the LGA model could be abandoned and the equilibria and collision matrices could be constructed independently to better suit numerical applications [23].

The simplest lattice Boltzmann equation (LBE) is the lattice BGK equation, based on a single-relaxation-time approximation [14]. Due to its extreme simplicity, the lattice BGK (LBGK) equation [8,9] has become the most popular lattice Boltzmann model in spite of its well known deficiencies.

The multiple-relaxation-time (MRT) lattice Boltzmann equation was also developed at the same time [20]. The MRT lattice Boltzmann equation (also referred to as the generalized lattice Boltzmann equation (GLBE) or the moment method) overcomes some obvious defects of the LBGK model, such as fixed Prandtl number ($Pr = 1$ for the BGK model) and fixed ratio between the kinematical and bulk viscosities. The MRT lattice Boltzmann equation has been persistently pursued, and much progress has been made. Successes include: formulation of optimal boundary conditions [24,25], interface conditions in multi-phase flows [26] and free surfaces [27], thermal [28] and viscoelastic models [29,30], models with reduced lattice symmetries [31,32], and improvement of numerical stability [33]. It should be stressed that most of the above results cannot be obtained with the LBGK method for most practical applications (RLBE schemes could be about 15% slower than their LBGK counterparts in terms of site-updating rate). Recently, it was shown that the multiple-relaxation-time LBE models are much more stable than their LBGK counterparts [33], because the different relaxation times can be individually tuned to achieve *optimal* stability.

The general RLBE model has three components. The first component is discrete phase space defined by a regular lattice in D dimensions together with a set of judiciously chosen discrete velocities e_α connecting each lattice site to some of its neighbors. The fundamental object in the theory is the set of velocity distribution functions f_α defined on each node r_i of the lattice. The second component includes a collision matrix S and $(N+1)$ equilibrium distribution functions $f_\alpha^{(eq)}$. The equilibrium distribution functions are functions of the local conserved quantities. The third component is the evolution equation in discrete time $t_n = n \cdot \delta_t$, $n = 0, 1, \ldots$,

$$\left| f\left(r_i + e_\alpha \delta_t, t + \delta_t\right)\right\rangle - \left| f\left(r_i, t\right)\right\rangle = -S\left[\left| f\left(r_i, t\right)\right\rangle - \left| f^{(eq)}\left(r_i, t\right)\right\rangle\right] \qquad (12)$$

The collision process is naturally accomplished in the space spanned by the eigenvectors of the collision matrix, the corresponding eigenvalues being the inverse of their relaxation time towards their equilibria. The $(N+1)$ eigenvalues of S are all between 0 and 2 as to maintain linear stability and the separation of scales, which means that the relaxation times of non-conserved quantities are much faster than the hydrodynamic time scales. The LBGK models are special cases in which the $(N+1)$ relaxation times are all equal, and the collision matrix $S = \omega \cdot I$, where I is the identity matrix, $\omega = 1/\tau$, and τ ($> \frac{1}{2}$) is the single relaxation time of the model.

The RLBE formulation has two important consequences: First, one has the maximum number of adjustable relaxation times, one for each class of kinetic modes invariant under the symmetry group of the underlying lattice. Second, one has maximum freedom in the construction of the equilibrium functions of the non-conserved moments. One immediate result of using the RLBE instead of the LBGK model is a significant improvement in numerical stability [33].

3.5.1   Optimization of the model and connection to the BGK LBE model

Among seven adjustable parameters (c_i, α_i, and γ_i) in the equilibrium values of the moments in the model (see Eq. 11), so far only five of these parameters have been fixed by enforcing the model to satisfy certain basic physics as shown in the preceding analysis: $c_1 = -2$, $\alpha_2 = -8$, $\gamma_1 = \gamma_3 = 2/3$, and $\gamma_2 = 18$. These parameter values are the optimal choice in the sense that they yield the desiderable properties (isotropy, Galilean invariance, etc.) to the highest possible order in wave vector k. It should be stressed that the constraints imposed by isotropy and Galilean invariance are beyond the conservation constraints –models with only conservation

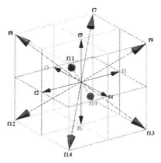

Figure 7. Discrete velocities of the D3Q15 model on a three-dimensional square lattice. *(Figure is taken from Freundiger* [34]).

constraints would not necessarily be isotropic a Galilean invariant in general, as observed in some newly proposed LBE models for non-ideal gases. Two other parameters, α_3 and γ_4, remain adjustable. In addition, there are six relaxation parameters s_i in the model as opposed to one in the LBE BGK model. Two of them, s_2 and s_8, determine the bulk and the shear viscosities, respectively. Also, because $c_1 = -2$, $s_9 = s_8$ [this reference]. The remaining three relaxation parameters, s_3, s_5 and s_7, can be adjusted without having any effect on the transport coefficients in the order of k^2. However, they do have effects in higher-order terms. Therefore, one can keep values of these three relaxation parameters only slightly larger than 1 (no severe over-relaxation effects are produced by these modes) so that the corresponding kinetic modes are well separated from those modes more directly affecting hydrodynamic transport.

It is interesting to note that the present model degenerates to the BGK LBE model [8,9] if we use a single relaxation parameter for all the modes, *i.e.*, $s_\alpha = 1/\tau$, and choose $\alpha_3 = 4$ and $\gamma_4 = -18$.

Therefore, in the BGK LBE model, all the modes relax with exactly the same relaxation parameter so there is no separation in time scales among the kinetic modes. This may severely affect the dynamics and the stability of the system, due to the coupling among these modes.

3.6   Lattice Boltzmann model in Three Dimensions

Each point on a unit cubic lattice space has six nearest neighbours, $(\pm1,0,0)$, $(0,\pm1,0)$, and $(0,0,\pm1)$, twelve next nearest neighbours, $(\pm1,\pm1,0)$, $(\pm1,0,\pm1)$, and $(0,\pm1,\pm1)$, and eight third nearest neighbours, $(\pm1,\pm1,\pm1)$. Elementary discrete velocity sets for lattice Boltzmann models in three dimensions are constructed from the set of twenty-six vectors pointing from the origin to the above neighbours and the zero vector $(0,0,0)$. The twenty-seven velocities are usually grouped into four subsets labelled by their squared modulus, 0,1,2, and 3. We also use the notation DdQq for the q-velocity model in d-dimensional space in what follows [9]. The fifteen discrete velocities in the D3Q15 (see Figure 7) model are:

$$e_\alpha = \begin{cases} (0,0,0), & \alpha = 0, \\ (\pm1,0,0),(0,\pm1,0),(0,0,\pm1), & \alpha = 1,2,...,6, \\ (\pm1,\pm1,\pm1) & \alpha = 7,8,...,14. \end{cases} \tag{13}$$

The components of the corresponding 15 orthogonal basis vectors are given by:

$$|m_0\rangle_\alpha = \|e_\alpha\|^0,$$

$$|m_1\rangle_\alpha = \|e_\alpha\|^2 - 2,$$

$$|m_2\rangle_\alpha = \frac{1}{2}\left(15\|e_\alpha\|^4 - 55\|e_\alpha\|^2 + 32\right)$$

$$|m_3\rangle_\alpha = e_{\alpha\,x},$$

$$|m_4\rangle_\alpha = \frac{1}{2}\left(5\|e_\alpha\|^2 - 13\right)e_{\alpha\,x},$$

$$|m_5\rangle_\alpha = e_{\alpha\,y},$$

$$|m_6\rangle_\alpha = \frac{1}{2}\left(5\|e_\alpha\|^2 - 13\right)e_{\alpha\,y},$$

$$|m_7\rangle_\alpha = e_{\alpha\,z},$$

$$|m_8\rangle_\alpha = \frac{1}{2}\left(5\|e_\alpha\|^2 - 13\right)e_{\alpha\,z},$$

$$|m_9\rangle_\alpha = 3e_{\alpha\,x}^2 - \|e_\alpha\|^2,$$

$$|m_{10}\rangle_\alpha = e_{\alpha\,y}^2 - e_{\alpha\,z}^2,$$

$$|m_{11}\rangle_\alpha = e_{\alpha\,x}e_{\alpha\,y},$$

$$|m_{12}\rangle_\alpha = e_{\alpha\,y}e_{\alpha\,z},$$

$$|m_{13}\rangle_\alpha = e_{\alpha\,x}e_{\alpha\,z},$$

$$|m_{14}\rangle_\alpha = e_{\alpha\,x}e_{\alpha\,y}e_{\alpha\,z}.$$

$$\alpha \in \{0,1,..., 14\} \tag{14}$$

where $\|e_\alpha\| = (e_{\alpha x}^2 + e_{\alpha y}^2 + e_{\alpha z}^2)^{\frac{1}{2}}$ and $\|e_0\|^0 = 1$. The corresponding fifteen moments are:

$$|m\rangle = \left(\rho,e,\varepsilon,j_x,q_x,j_y,q_y,j_z,q_z,3p_{xx},p_{ww},p_{xy},p_{yz},p_{zz},m_{xyz}\right)^T. \tag{15}$$

The collision matrix S in moment space **M** is the diagonal matrix

$$S \equiv diag\left(0,s_1,s_2,0,s_4,0,s_4,0,s_4,s_9,s_9,s_9,s_{11},s_{11},s_{14}\right), \tag{16}$$

zeros corresponding to conserved moments in the chosen order. And the matrix M is then given by

$$M = \begin{vmatrix}
1 & 1 & 1 & 1 & 1 & 1 & 1 & 1 & 1 & 1 & 1 & 1 & 1 & 1 & 1 \\
-2 & -1 & -1 & -1 & -1 & -1 & -1 & 1 & 1 & 1 & 1 & 1 & 1 & 1 & 1 \\
16 & -4 & -4 & -4 & -4 & -4 & -4 & 1 & 1 & 1 & 1 & 1 & 1 & 1 & 1 \\
0 & 1 & -1 & 0 & 0 & 0 & 0 & 1 & -1 & 1 & -1 & 1 & -1 & 1 & -1 \\
0 & -4 & 4 & 0 & 0 & 0 & 0 & 1 & -1 & 1 & -1 & 1 & -1 & 1 & -1 \\
0 & 0 & 0 & 1 & -1 & 0 & 0 & 1 & 1 & -1 & -1 & 1 & 1 & -1 & -1 \\
0 & 0 & 0 & -4 & 4 & 0 & 0 & 1 & 1 & -1 & -1 & 1 & 1 & -1 & -1 \\
0 & 0 & 0 & 0 & 0 & 1 & -1 & 1 & 1 & 1 & 1 & -1 & -1 & -1 & -1 \\
0 & 0 & 0 & 0 & 0 & -4 & 4 & 1 & 1 & 1 & 1 & -1 & -1 & -1 & -1 \\
0 & 2 & 2 & -1 & -1 & -1 & -1 & 0 & 0 & 0 & 0 & 0 & 0 & 0 & 0 \\
0 & 0 & 0 & 1 & 1 & -1 & -1 & 0 & 0 & 0 & 0 & 0 & 0 & 0 & 0 \\
0 & 0 & 0 & 0 & 0 & 0 & 0 & 1 & -1 & -1 & 1 & 1 & 1 & -1 & 1 \\
0 & 0 & 0 & 0 & 0 & 0 & 0 & 1 & 1 & -1 & -1 & -1 & -1 & -1 & 1 \\
0 & 0 & 0 & 0 & 0 & 0 & 0 & 1 & -1 & 1 & -1 & -1 & 1 & -1 & 1 \\
0 & 0 & 0 & 0 & 0 & 0 & 0 & 1 & -1 & -1 & 1 & -1 & 1 & 1 & -1
\end{vmatrix} \tag{17}$$

The equilibria of the kinetic moments as functions of $\rho^{(eq)} = \rho$ and $j^{(eq)} = j$ up to second-order are given by

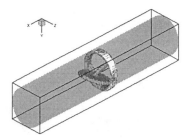

Figure 8. View of the valve geometry used for simulation, i.e. HK Hall-Kaster, Medtronic Hall. *(Figure is taken from Cárdenes et al. [35]).*

$$e^{(eq)} = -\rho + \frac{1}{\rho_0} j \cdot j = -\rho + \frac{1}{\rho_0}\left(j_x^2 + j_y^2 + j_z^2\right),$$

(18)

$$\varepsilon^{(eq)} = -\rho, \qquad q_x^{(eq)} = -\frac{7}{3} j_x,$$

$$q_y^{(eq)} = -\frac{7}{3} j_y, \qquad q_z^{(eq)} = -\frac{7}{3} j_z,$$

$$p_{xx}^{(eq)} = \frac{1}{3\rho_0}\left[2 j_x^2 - \left(j_y^2 + j_z^2\right)\right], \qquad p_{ww}^{(eq)} = \frac{1}{\rho_0}\left[j_y^2 - j_z^2\right],$$

$$p_{xy}^{(eq)} = \frac{1}{\rho_0} j_x j_y, \qquad p_{yz}^{(eq)} = \frac{1}{\rho_0} j_y j_z,$$

$$p_{xz}^{(eq)} = \frac{1}{\rho_0} j_x j_z, \qquad m_{xyz}^{(eq)} = 0.$$

The constant ρ_0 is the mean density in the system and is usually set to be unity in simulations. The approximation of $1/\rho \approx 1/\rho_0$ is used to reduce compressibility effects in the model.

4   Simulations

4.1   *Three-dimensional transient physiological flows in fixed geometries by lattice Boltzmann BGK method*

In order to simulate and compare the behavior of the different MHV's, the following considerations are taken: the fluid dynamic problem is simplified by assuming that the valves are fixed and completely open. Three different MHV models are used: a Starr-Edwards[TM] caged ball, a HK Hall-Kaster Medtronic Hall and a St. Jude Medical valve. The valves are placed in a control volume with the following dimensions: length = 100 *mm* and the diameter of the tube = 20 *mm* (see Figure 8). A fully developed parabolic velocity profile at inlet and outlet is taken as boundary condition with a mean velocity of 66 *mm/s*. It was taken from the time-

Figure 9.   Surf plots of shear stress ratio in the yz-plane $| (\tau_{xy} - \tau_{xyMAX}) / \tau_{xyMAX} |$ for a Starr-Edwards[TM] caged ball heart valve.

dependant cardiac cycle in order to get a low Reynolds number in order to simulate laminar non-transient Newtonian flow. The mathematical approach used for the simulation of three-dimensional transient physiological flows is the Lattice–Boltzmann BGK method.

From Figure 9 to Figure 14 show the shear stress ratio $| (\tau_{xy} - \tau_{xyMAX}) / \tau_{xyMAX} |$ developed by the flow that pass through the artificial valves and the streamlines. τ_{xyMAX} is the maximum value obtained for the shear stress calculated from the three valves. In the Figure 10 it can be noted the vortex formation when the flow finally get pass through the ball; the streamlines in the Figure 12 and Figure 14 not show significance alteration of the flow from downstream to upstream.

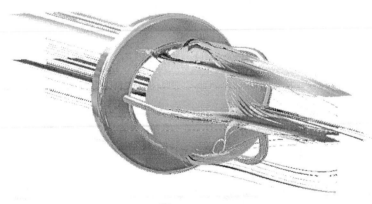

Figure 10.   Streamlines for a Starr-Edwards[TM] caged ball heart valve.

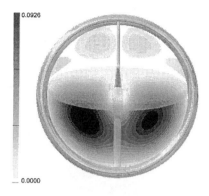

Figure 11. Surf plots of shear stress ratio in the yz-plane $| (\tau_{xy} - \tau_{xyMAX}) / \tau_{xyMAX} |$ for a HK Hall-Kaster Medtronic Hall valve.

4.2   Two-dimensional transient physiological flows with moving boundaries by general lattice Boltzmann equation

The thromboembolic potential effects of MHV implantation were studied by a numerical simulation of time dependent blood flow through the valve. The physiological conditions after MHV implantation were modeled in detail (Figure 15); the natural geometry of the aortic sinuses was modified to conform with aortotomy (native valve excision) and the MHV suturing (sub annular) onto the aortic annulus, which takes place within a couple of weeks after valve implantation [36,37]. Geometry was based on Atlas of cardiothoracic surgery [38]. The evaluated models were the valve HK (Hall-Kaster; Medtronic Hall) and the St. Jude Medical.

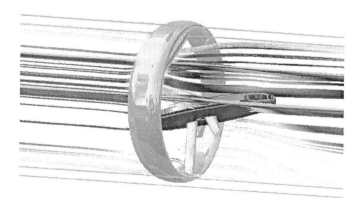

Figure 12.   Streamlines for a HK Hall-Kaster Medtronic Hall valve.

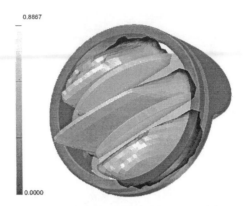

0.8867

0.0000

Figure 13. Surf plots of shear stress ratio in the yz-plane $| (\tau_{xy} - \tau_{xy\text{MAX}}) / \tau_{xy\text{MAX}} |$ for a St. Jude Medical valve.

4.2.1   Unsteady Flow: 2D Womersley Flow

Dimensionless numbers are very useful in characterizing mechanical behavior because their magnitude can often be interpreted as the relative importance of competing forces that will influence mechanical behavior in different ways. One dimensionless number, the Womersley number (Wo), is sometimes used to describe the unsteady nature of fluid flow in response to an unsteady pressure gradient, *i.e.,* whether the resulting fluid flow is quasi-steady or not. Fluids surround organisms which themselves contain fluid compartments; the behaviors exhibited by these biologically –important fluids (*e.g.* air, water, or blood) are physiologically significant because they will determine, to a large extent, the rates of mass and heat exchange and the force production between an organism and its environment or between different parts of an organism.

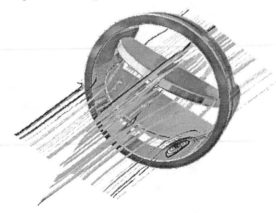

Figure 14. Streamlines for a St. Jude Medical valve.

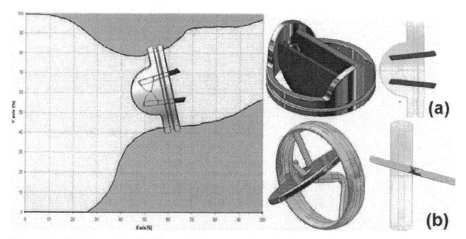

Figure 15. Model studied of MHV geometry after implantation in the aortic position. (a) for a St. Jude Medical model, and (b) for HK Medtronic Hall.

The *Wo* number is defined by,

$$Wo = \frac{L}{2}\sqrt{\frac{n}{v}} \qquad (19)$$

where L is the characteristic length in m; n is the frequency of the unsteady flow or movements in *radians/s* ($n = 2\pi f$ where f is frequency in *cycles/s*) and v is the kinematical viscosity of the fluid in m^2/s. At the present time this dimensionless parameter is used almost exclusively for cases of flow inside circular cylinders, at least in the biological literature [39,40,41,42], presumably because this is the geometry for which a mathematical foundation has been laid in an explicit biological context. Four decades ago, such foundation was laid in a study on flow in mammalian blood vessels [43]; this flow is unsteady because of the rhythmic nature of pressure applied by a beating heart. Womersley identified a dimensionless parameter group (Eq. 19, L = internal diameter of the cylindrical vessel), later

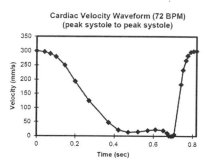

Figure 16. Transient inlet velocity waveform at the ventricle side *(Figure is taken from Bluestein et al.* [37]).

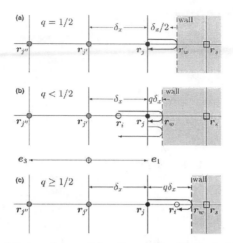

Figure 17. Illustration of the boundary conditions for a rigid wall located arbitrarily between two grid sites in one dimension. The thin solid lines are the grid lines, the dashed line is the boundary location situated arbitrarily between two grids. Shaded disks are the fluid nodes, and the disks (●) are the fluid nodes next to boundary. Circles (○) are located in the fluid region but not on grid nodes. The square boxes (□) are within the non-fluid region. The thick arrows represent the trajectory of a particle interacting with the wall. The distribution functions at the locations indicated by disks are used to interpolate the distribution function at the location marked by the circles (○). (a) $q \equiv |r_w - r_j| / \delta_x = \frac{1}{2}$. This is the perfect bounce-back condition – no interpolations needed. (b) $q < \frac{1}{2}$. (c) $q \geq \frac{1}{2}$. *(Figure is taken from Lallemand et al.* [44]*).*

named after him, useful in indicating a dichotomy in fluid behaviour: when $Wo<1$, the fluid behaves in a "quasi-steady" manner, while for $Wo>1$, the behaviour of the fluid deviates more and more from quasi-steady behaviour. Note that this *standard* terminology can be confusing: *"quasi-steady"* does not mean *"approximately steady"* (*i.e.*, not changing very much in time). Rather, "quasi-steady" means that at any time, the instantaneous flow rate is determined by the instantaneous pressure gradient. Thus, quasi-steady flow can actually oscillate more vigorously than non-quasi-steady flow simply because it will keep up with a rapidly changing pressure gradient. Womersley's contribution was evaluated by Mc-Donald [45], who pointed out that there were "no theoretical innovations but it is original in that it is put in a form that can be computed easily" (this case of computation assumes familiarity with Bessel functions and Fourier series). Parallel but separate mathematical treatments at that same time in the engineering literature may be accessed by consulting Schlichting [46] or White [47].

In many biological applications an investigator may need only a qualitative indicator of fluid behaviour. Because of its appearance in the Navier-Stokes equations, Wo is such a qualitative indicator for unsteady flow behaviour. Before associations may be made between unsteady fluid behaviours and magnitudes of Wo, the appropriate foundation must be laid for the relevant physical setting (which will include the geometry and Reynolds number range).

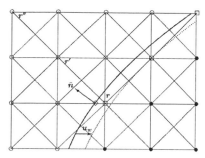

Figure 18. Illustration of a moving boundary with velocity u_w. The circles (o) and discs (•) denote the fluid and solid nodes, respectively. The squares (□) denote the nodes becoming fluid nodes from the solid nodes at one time $\delta_t = 1$. The solid and dotted curves are the wall boundary at time t and $t + \delta_t$, respectively. *(Figure is taken from Lallemand et al.* [44]*).*

For the simulation, blood was modelled as a fluid with density of $\rho=1.050$ g/cm^3 and $v= 3.4$ mm^2/s. The numerical simulations were conducted in the numerical mesh with cardiac output of 5.5 l/min and heart rate of 72 bpm (820 ms, Wo= 14.9). The temporal value of the velocity specified at the inlet was taken from a typical physiological waveform [48] (Figure 16). Additionally, 4 diameters exit length were used to eliminate outlet effects. An aortic diameter of 20 *mm* was considered.

First, a set of simulations of measuring velocity profiles across the geometry at different times were conducted. In the simulations, the period is $T=10^5$ time steps. The initial state of velocity field in the system is set to zero. The velocity field evolve for $10T$ initial steps to ensure independence from initial effects using the convergence criterion

$$\sum_i \frac{\|u(x_i,t+T)-u(x_i,t)\|}{\|u(x_i,t+T)\|} \leq 10^7 \qquad (20)$$

where the summation is over the entire system.

4.2.2   Lattice Boltzmann Method for moving boundaries

An area of interest within LBE method is the form in which the boundary conditions are considered. Particularly, the method of *bounce-back* is one of the most used and studied, which mimics the particle-boundary interaction for no-slip boundary condition by reversing the momentum of the particle colliding with an impenetrable and rigid wall.

The *bounce-back* condition, additionally to being of easy implementation, is precise and appropriate for very simple geometries boundaries done with rectilinear segments. Now, to consider complex geometries with arbitrary curvatures, two strategies in LBE method have been applied. The first strategy used, is to create a body-fitted mesh and calculate interpolations throughout entire mesh in addition to the advection process, since the computational mesh does not overlap with the underlying Cartesian lattice; then, the bounce-back condition to the solid contour nodes is applied.

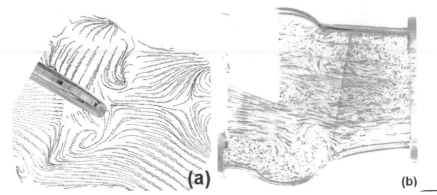

Figure 19. Velocity vector plot: (a) at $0.93T$ ($t = 0.744\,s$) in a one leaflet MHV. Comparison with a PIV velocity vector field [49] (b) in the aortic root.

The second strategy maintains the regular Cartesian mesh and applies interpolations to locate the contour line in the mesh, and the bounce-back conditions are applied in the obtained position of the contour, which may be out of the Cartesian points grid. These methods are used for fixed objects in fluid simulations.

For the simulations with moving and curved boundaries, the Lallemand & Luo proposal was used. This proposal is a simple extension of the treatment for a curved boundary proposed by Bouzidi *et al* [50]. The Figure17 illustrates the simplicity of this boundary condition under a bi-dimensional example.

It shows a wall located in an arbitrary position r_w, between two lattice sites, r_j and r_s (last point located within the solid region). The variable q is defined as the fraction in the fluid region of a grid spacing $|r_s - r_j|$ intersected by the wall. For case $q \equiv |r_w - r_j|/\delta_x < \frac{1}{2}$ (see Figure17 b) the scheme depicts the detail of the particle r_j for a time t after the bounce-back collision. The distribution function of this node with characteristics of *FLOW* and velocity pointing to the wall (e_l in the Figure) can end at the point r_i located at a distance $(1\text{-}2q)\ \delta_x$ from the point r_j of the grid. As the point r_i is not a LB grid point, the value of f_3 for r_j needs to be reconstructed (Noticing that the point f_l that start from r_i would become f_3 at the grid point r_j after the bounce-back collision).

The value of f_l in r_i is calculated by a quadratic interpolation using the f_l values in the grid points: j, j' and j''. In a similar case, for the case $q \geq \frac{1}{2}$ (Figure 17 c) reconstruct f_l through the f_3 in r_j by quadratic interpolation, relating $f_3(r_i)$ that is equal to $f_l(r_j)$ before the bounce-back collision and f_3 at the nodes after the collision and propagation: $f_3(r_j')$ and $f_3(r_j'')$. In any case the interpolations are applied differently for both cases:

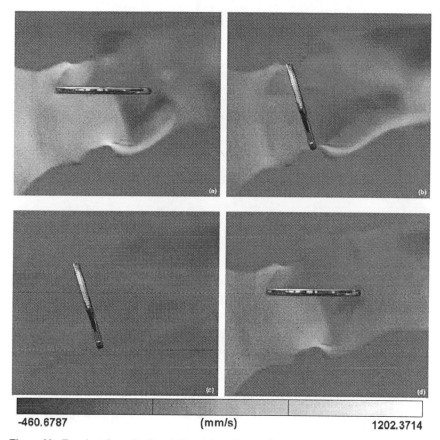

-460.6787   **(mm/s)**   1202.3714

Figure 20. Transient dynamics from left to right and top to bottom: $0T$ ($t_a = 0.0\ s$), $0.3T$ ($t_b = 0.24\ s$), $0.87T$ ($t_c = 0.696\ s$) and $0.93T$ ($t_d = 0.744\ s$) velocity magnitude $|u_x|$.

- For $q < \frac{1}{2}$, interpolation is done before the propagation and bounce-back collision.
- For $q \geq \frac{1}{2}$, interpolation is done after the propagation and bounce-back collision.

In order to extend these boundary conditions to solid walls in movement (see Figure 18), the velocity of the moving boundary, u_w, must not be too fast when compared with the sound speed c_s in the system, and the following problem must be solved additionally: when a lattice point moves outside the solid region towards the fluid region ($\square$ in the Figure 18), it will require the reconstruction of the values associated to his fluid distribution function, unknown for the previous iteration. The method uses a second order extrapolation to compute the unknown distribution functions along the direction of a chosen discrete velocity e_a which maximizes the

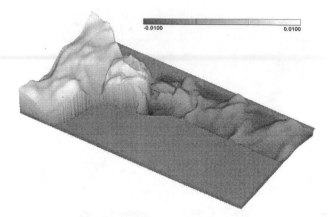

Figure 21. Contours of drop pressure in the xy-plane: $0.93T$ $(t = 0.744\ s)$ $\Delta p/p_{ref}$.

quantity $\hat{n} \cdot e_\alpha$, where $\hat{n}$ is the out-normal vector of the wall at the point ($\Diamond$ in the Figure 18 through which the node moves to the fluid region. For example, the unknown distribution function $\{f_\alpha(r)\}$ at node r, showed in Figure 18, can be obtained by the following extrapolation formula:

$$f_\alpha(r) = 3f_\alpha(r') - 3f_\alpha(r'') + f_\alpha(r'' + e_6) \tag{21}$$

Obviously this is not the unique method for extrapolation. One could, for example, calculate the equilibrium distribution functions at r by using the velocity u_w of the moving boundary and the averaged density in the system ρ_0 or the local averaged density, and use the equilibrium distribution functions for the unknown distribution functions. The results of these procedures are similar.

As depicted in Figure 18, the momentum transfer occurred near the boundary along the direction of e_α is equal to

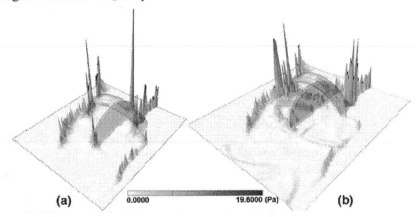

Figure 22. Surf plots of shear stress in the xy-plane: $0.93T$ $(t=0.744\ s)$ $|\tau_{xy}|$. (a) Shows results with a leaflet in movement (moving boundary), and (b) with a fixed leaflet (top opening angle).

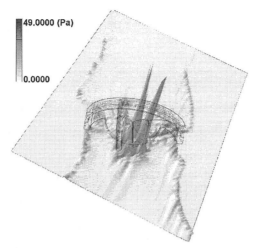

Figure 23. Surf plots of shear stress in the *xy*-plane: 0.20*T* (*t*=0.16 *s*) | τ$_{xy}$ |. Results with leaflets in movement (moving boundary).

$$\delta p_\alpha = e_\alpha \left[f_\alpha(r_w, t) + f_{\bar{\alpha}}(r_j, t) \right] \tag{22}$$

where $f_{\bar{\alpha}}$ is the distribution function of the velocity $e_{\bar{\alpha}} \equiv -e_\alpha$. The above formula gives the momentum flux through any boundary normal to e_α located between the point r_w and the point r_j in Figure 18.

Basically the dynamics of the rigid body is conformed by two basic movements: Translation and rotation [51]. For convenience the acceleration vector is defined at the Center of Mass coords, called *G*.

The general equations are:

TRANSLATION
$$\sum_{i=1}^{n} \vec{F}_i = m \cdot \vec{a}_G \tag{23}$$

ROTATION
$$\vec{M}_G = \left(\frac{d\vec{L}_G}{dt} \right)_{xyz} + \vec{\Omega} \times \vec{L}_G \tag{24}$$

The moments and products of inertia are invariant in the time. Then we have:
$$M = I \cdot \dot{\omega} \tag{25}$$
and
$$\omega(t + \delta t) = \omega(t) + \dot{\omega}(t) \delta t \tag{26}$$

Knowing the vector direction, the velocity and the acceleration module is calculated by:

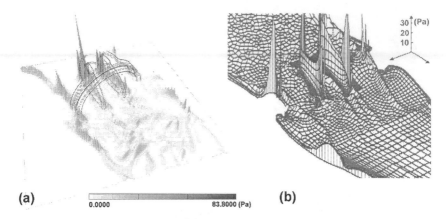

(a)

0.0000   83.8000 (Pa)   **(b)**

Figure 24. Surf plots of shear stress in the *xy*-plane: $0.25T$ ($t=0.20\ s$) $|\ \tau_{xy}\ |$. We show results with fixed leaflets (top opening angle). (a) LBE simulation, and (b) FEM simulation using the FIDAP CFD package *(Figure is taken from Bluestein et al.* [37]*).*

$$\theta(t+\delta t)=\theta(t)+\omega(t)\delta t+\frac{\dot{\omega}(t)\delta t^{2}}{2} \qquad (27)$$

For the analyzed valve the translation movement was not consider. In the case of SJM valve, the movement of each leaflet is studied separately.

Figure 19, shows streamlines of monoleaflet MHV at times $t = 0.93T$. Lim *et al.* [49] developed a MHV testing system suitable for particle image velocimetry (PIV). This shows similar results corresponding to peak systole phase. We found a slightly smaller recirculating region in the sinus cavity and a more pronounced wake region along the valve centreline with low velocities.

The Figure 20 shows surface plots of u_x in the *xy*-plane for the same valve at different times on the cardiac velocity waveform. In Figure 21, drop pressure contours are plotted. Here, the pressures are nondimensionalized by p_{ref}, where p_{ref} is the average pressure in the aorta. The pressure has also very steep gradient in this region, reflecting the rapid local dynamics of the flow.

An important issue for valve hemodynamics analysis is shear stress levels in the fluid. Although the aortic sinuses were partially flattened as a result of the aortotomy, the combination of the collar stenosis and the sinuses still generated typical aortic sinus vortices. The shear layers formed in the interface between the jets and recirculation zones produce elevated flow stresses, which contribute to platelet activation.

Figure 22 and Figure 23 show the absolute value of shear stress levels in the fluid for the two MHV models. Figure 24 shows absolute values of shear stress levels for the SJM considering the manufacturers predefined maximum leaflet opening angle fixed. Bluestein *et al.* [37] shows similar results at the same time (200 *ms* after peak systole). The leaflets appear concealed between the steep shear layers.

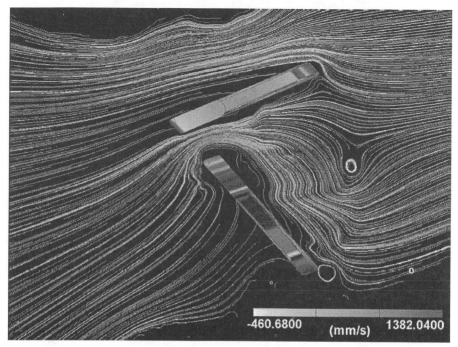

Figure 25. Streamlines for a Saint Jude Medical valve: $0.20T$ ($t=0.16\ s$) considering moving boundary.

The bileaflet MHV simulation was more complicated, because the model turned out to be more sensitive to the fluctuations induced by the moving boundary conditions.

Flow patterns that contribute to the cardioembolic potential of this MHV include areas of elevated shear stresses in the interfaces between the typical three-jet flow generated through a MHV holes, and the ensuing recirculation zones and vortices [37].

To ensure small *Mach* and *Knudsen* numbers, significant fluctuations in the distribution functions around the moving solid alter the results and influence the tumbling of the leaflets. Some partial results for the Saint Jude Medical valve are shown for comparison (see Figure 25).

Investigations to reduce the aforementioned errors of the moving boundary method used in this work are been studied [52] and are expected to improve future simulations.

5   Conclusions and Outlook

The goal of this study was to test the suitability of LB method for two and three-dimensional bioengineering pulsatile flow problems. Preliminary LBE simulations of blood flow with moving boundaries have qualitatively reproduced experimental values. Nevertheless, the present use of a uniform grid decreases the possibilities of the present algorithm to model the detailed influence of leaflets with varying thicknesses and other geometric details which are important for the accurate prediction of local peak stress fields in the moment of valve closure. The reason is that there is a spatial fluctuation in the force depending on the location of the boundary relative to the mesh: there are grid points moving from solid region to fluid region (and *vise versa*). When a grid point just emerges from the solid region into the fluid region, the distribution functions must be constructed on this point. The grid points moving in and out of the fluid region to and from the solid region have an immediate effect on the momentum transfer at the boundary near their neighborhoods.

Another problem is to find optimal interpolation schemes to minimize mass conservation errors near the boundary.

While the two-dimensional approximation is a viable model for the basic dynamics of transient blood flow passing through a MHV, future work is under way to analyze three-dimensional flow dynamics.

Acknowledgments

The authors wish to thank Dr. Manfred Krafczyk from the Institute for Computer Applications in Civil Engineering (Technical University of Braunschweig – Germany) for very useful discussions and help in lattice Boltzmann theory formulation. This work was supported by the Council for Scientific Humanistic Development (C.D.C.H. - Venezuela) in the framework of the Cardioma Project and the Venezuelan Health Ministry, which are gratefully acknowledged.

References

1. O'Malley C.D. and Saunders C.M., *Leonardo da Vinci on the Human Body,* Henry Schuman, N.Y. (1952).
2. Hufnagel C.A. and Harvey W.P., *Surgical correction of aortic insufficiency: Preliminary report,* Bull, Georgetown Univ. Med. Ctr., **6** (1953).
3. Wolfram S. ed., *Theory and Applications of Cellular Automata*, World Scientific (1986).
4. Wolfram S., Cellular automata as models of complexity, *Nature* **311** (1984) pp. 419.
5. Packard N. and Wolfram S., Two-dimensional cellular automata, *J. Stat. Phys.* **38** (1985) pp. 901.
6. Tritton D.J., *Physical Fluid Dynamics* Van Nostrand (1977).

7. McNamara G. and Zanetti G., Use of the Boltzmann equation to simulate lattice-gas automata, *Phys. Rev. Lett.* **61** (1988) pp. 2332–2335.
8. Chen H., Chen S., Matthäus W.H., Recovery of the Navier-Stokes equations using a lattice-gas Boltzmann approach, *Phys. Rev. A* **45** (1992) pp. R5339–R5342.
9. Qian Y.H., d'Humières D., Lallemand P., Lattice BGK models for Navier-Stokes eqution *Europhys. Lett.* **17** (1992) pp. 479–484.
10. Junk M., *Discrete Modelling and Discrete Algorithms in Continuum Mechanics*, T. Sonar and I. Thomas, eds. , Logos, Berlin (2001).
11. Frisch U., Hasslacher B., Pomeau Y., Lattice-gas automata for the Navier-Stokes equation, *Phys. Rev. Lett.* **56** (1986) pp. 1505–1508.
12. d'Humières D., Lallemand P., Frisch U., Lattice gas models for 3d hydrodynamics, *Europhys. Lett.* **2**, (1986) pp. 291–297.
13. Higuera F.J., Jiménez J., Boltzmann approach to lattice gas simulations *Europhys. Lett.* **9** (1989) pp. 663–668.
14. Bhatnagar P.L., Gross E.P., Krook M., A model for collision processes in gases. I. Small amplitude processes in charged and neutral one-component systems, *Phys. Rev.* **94** (1954) pp. 511–524.
15. Abe T., Derivation of the lattice Boltzmann method by means of the discrete ordinate method for the Boltzmann equation, *J. Comput. Phys.* **131** (1997) pp. 241–246.
16. He X., –S. Luo L., Theory of the lattice Boltzmann method: From the Boltzmann equation to the lattice Boltzmann equation, *Phys. Rev. E* **56** (1997) pp. 6811–6817.
17. Junk M., On the construction of discrete equilibrium distributions for kinetic schemes, *ITWM Report* **14**, Institut für Techno- und Wirtschaftsmathematik, Kaiserslautern (1999) pp. 947–968.
18. Luo L. –S., Unified theory of lattice Boltzmann models for nonideal gases, *Phys. Rev. Lett.* **81** (1998) pp. 1618–1621.
19. Luo L. –S., Theory of the lattice Boltzmann method: lattice Boltzmann models for nonideal gases, *Phys. Rev. E* **62** (2000) pp. 4982–4996.
20. d'Humières D., Generalized lattice Boltzmann equations, *Rarefied Gas Dynamics: Theory and Simulations, Progress in Astronautics and Aeronautics*, AIAA **159** (1992) pp. 12.
21. Hirschfelder J.O., Curtiss C.F., Bird R.B., *Molecular Theory of Gases and Liquids*, Wiley, N.Y. (1954).
22. Harris S., *An Introduction to the Theory of the Boltzmann Equation*, Holt, Rinehart and Winston, N.Y. (1971).
23. Higuera F.J., Succi S, Benzi R., Lattice gas dynamics with enhanced collisions, *Europhys. Lett.* **9** (1989) pp. 345–349.
24. Ginzbourg I., Adler P.M., Boundary flow condition analysis for the three-dimensional lattice Boltzmann model, *J. Phys. II France* **4** (1994) pp. 191–214.

25. Ladd A.J.C., Numerical simulations of particulate suspensions via a discretized Boltzmann equation. Part 1. Theoretical foundation, *J. Fluid Mech.* **271** (1994) pp. 285–309.
26. Ginzbourg I., Adler P.M., Surface tension models with different viscosities, *Transport in Porous Media* **20** (1995) pp. 37–76.
27. Ginzbourg I., Steiner K., *Free surface lattice-Boltzmann method to model the filling of expanding cavities by Bingham Fluids*, These proceeding.
28. McNamara G.R., Garcia A.L., Alder B.J., Stabilization of thermal lattice Boltzmann models, *J. Stat. Phys.* **81** (1995) pp. 395-408.
29. Giraud L., D'Humières D., Lallemand P., A lattice Boltzmann model vor visco-elasticity, *Int. J. Mod. Phys.* **8** (1997) pp. 805–815.
30. Giraud L., D'Humières D., Lallemand P., A lattice Boltzmann model for Jeffreys viscoelastic fluid, *Europhys. Lett.* **42** (1998) pp. 625-630.
31. d'Humières D., Bouzidi M., Lallemand P., Thirteen-velocity three-dimensional lattice Boltzmann model, *Phys. Rev.* **63** (2001) pp. 066702-1–7.
32. Bouzidi M., d'Humières D., Lallemand P.,–S. Luo L., Lattice Boltzmann equation on a two-dimensional rectangular grid, *J. Comput. Phys.* **172** (2001) pp. 704–717.
33. Lallemand P., –S. Luo L., Theory of the lattice Boltzmann method : Dispersion, dissipation, isotropy, Galilean invariance, and stability, *Phys. Rev. E* **61** (2000) pp. 6546–6562.
34. Freudiger S., *Effiziente Datenstrukturen für Lattice-Boltzmann-Simulationen in der computergestützten Strömungsmechanik*, TU München (2001).
35. Cárdenes J., Estrada H., *Simulación y análisis computacional del flujo en válvulas artificiales del corazón a través del método de lattice Boltzmann*, Universidad Central de Venezuela (2002).
36. Deviri E., Sareli P., Wisenbaugh T., Cronje S.L., Obstruction of mechanical heart valve prostheses: clinical aspects and surgical management, *J. of the American College of Card.* **17** (1991) pp. 646–650.
37. Bluestein D., Li Y.M., Krukenkamp I.B., Free emboli formation in the wake of bi-leaflet mechanical heart valves and the effects of implantation techniques, *J. of Biomech.* **35** (2002) pp. 1533–1540.
38. Edmunds L.H., Norwood W.I., Low D.W., *Atlas of cardiothoracic surgery*, Lea&Febiger, Philadelphia-London (1990).
39. Pedley T.J., Schroter R.C., Sudlow M.F., *Bioengineering Aspects of the Lung*, West J.B., ed., Marcel Dekker Inc., N.Y. (1977).
40. Caro C.G., Pedley T.J., Schroter R.C., Seed W.A., *The Mechanics of the Circulation*, Oxford: Oxford University Press (1978).
41. Daniel T.L., Kingsolver J.G., Meyhofer E., Mechanical determinants of nectar feeding energetics in butterflies: muscle mechanics, feeding geometry and functional equivalence, *Oecologia* **79** (1989) pp. 66–75.

42. Vogel S., *Life in Moving Fluids: The Physical Biology of Flow*, 2nd ed., Princeton: Princeton University Press (1994).
43. Womersley J.R., Method for the calculation of velocity, rate of flow and viscous drag in arteries when the pressure gradient is known, *J. Physiol.* **127** (1955) pp. 553–563.
44. Lallemand P., –S. Luo L., Lattice Boltzmann method for moving boundaries, *J. of Comp. Phys.* **184** (2003) pp. 406–421.

45. McDonald D.A., The relation of pulsatile pressure to flow in arteries, *J. Physiol.* **127** (1955) pp. 533–552.
46. Schlichting H., *Boundary-Layer Theory*, 7th ed., McGraw-Hill, N.Y. (1979).
47. White F.M., *Viscous Fluid Flow.* 2nd ed., McGraw-Hill, N.Y. (1991).
48. Chaux A., Blanche C., Technical aspects of valvular replacement with the St. Jude prosthesis, *J. of Cardiovascular Surgery* **28** [4] (1987) p, 363.
49. Lim W.L., Chew Y.T., Chew T.C., Low H.T., Steady flow dynamics of prosthetic aortic heart valves: a comparative evaluation with PIV techniques, *J. of Biomech.* **31** (1998) pp. 411–421.
50. Bouzidi M., Firdaouss M., Lallemand P., Momentum transfer of a lattice-Boltzmann fluid with boundaries, *Phys. Fluids* **13** (2002) pp. 3452–3459.
51. Martínez J.J., *Mecánica newtoniana*, UPC ed. (2001).
52. Ginzbourg I., d'Humières D., Multi-reflection boundary conditions for lattice Boltzmann models, *ITWM Report* **38** (2002) available at http://www.itwm.uni-kl.de/.